Institutional Diversity and Sustainable Environmental Management

Scalar, Cultural, and Functional Perspectives

Editors

H.M. Tuihedur Rahman

Department of Agricultural and Resource Economics
College of Agriculture and Bioresources
University of Saskatchewan
Saskatoon, Canada

Ashlee-Ann Pigford

Alumni of the Department of Natural Resource Sciences
McGill University
Ste-Anne-de-Bellevue
Quebec, Canada

CRC Press
Taylor & Francis Group
Boca Raton London New York

CRC Press is an imprint of the
Taylor & Francis Group, an **informa** business

A SCIENCE PUBLISHERS BOOK

Cover Art by Stephanie Louise Aitken

First edition published 2025
by CRC Press
2385 NW Executive Center Drive, Suite 320, Boca Raton FL 33431

and by CRC Press
4 Park Square, Milton Park, Abingdon, Oxon, OX14 4RN

Library of Congress Cataloging-in-Publication Data (applied for)

ISBN: 978-0-367-48809-3 (hbk)
ISBN: 978-1-032-99548-9 (pbk)
ISBN: 978-1-003-04291-4 (ebk)

DOI: 10.1201/9781003042914

Typeset in Times New Roman
by Prime Publishing Services

Dedication

To the loving memory of Mrs. Fazila Begum,
mother of Dr. Rahman

Foreword

Institutions are the social structures affecting collective action toward a common purpose. From fields including political science, economics, and management to education, psychology, sociology, anthropology, and religion, the study of institutions draws on historical, rational choice, and organizational perspectives to understand human interactions and how they can be influenced. In applied environmental management and governance scholarship, studying institutions and their cross-scalar, cross-cultural, and cross-functional dimensions is essential. Institutions present some of the most exciting opportunities for, and significant barriers to, realizing the sustainable development objectives of society. Of policy interest is understanding how diverse institutionalized behaviours form, interact, and perpetuate in different decision-making scenarios, as well as the processes of deinstitutionalizing prevailing behaviours and institutionalizing new ones when deemed necessary. This invites multi-dimensional concepts such as trust, risk, power, and control to be considered. The chapters in this volume each make an important contribution to understanding the many roles being played by institutions in collective action behaviours. Together, they advance the conceptualization of institutional analysis in environmental and natural resource management systems. Spanning local to international level decision-making contexts and embracing the complexity of formal and informal institutional interactions in settings including Bangladesh, New Zealand, the Philippines, Jamaica, and the Arctic, the inter-organizational, inter-generational, and inter-cultural dimensions of institutional diversity are made clear. The importance of social justice, equity, democratic engagement, and social inclusion to ensure participation and representation are common themes throughout the book. The need for greater coordination and innovation is also highlighted, with case studies emphasizing the importance of flexibility, adaptability, and learning to help institutions remain relevant. The call for methodological and conceptual advances to realize a more comprehensive understanding of institutional diversity and polycentricity in environmental management is timely. I was very pleased to be invited by the Editors to write a Foreword to this collection, which includes contributions from many of my former students and postdoctoral trainees who have since continued their exploration of institutional diversity in different natural resource systems internationally. I share their commitment to better identifying and addressing inter-institutional gaps and interplays to ensure the equitability and sustainability of environmental management regimes. The rich case studies presented in this volume underscore the relevance of institutional work to diverse communities and the environments on which they depend. I recommend this book to anyone working on environmental and natural resource management, policy, and governance challenges.

Gordon M. Hickey
McGill University, Canada

Preface

The idea for this book emerged from a long-standing collaboration among a group of international scholars studying the impacts of institutional diversity on sustainable environmental management. Recent studies on environmental governance demonstrate that there is no 'one size fits all' institutional strategy for managing the changing and at times problematic contexts of interlinked social and environmental systems. Therefore, innovative and dynamic approaches for institutional development are required to better promote sustainability. These approaches need to foster coordination among diverse environmental institutions with differentiated social, political, cultural, and economic interests. However, a lack of empirical examples describing the properties of diverse institutions and their interactions impedes our capacity to understand and design innovative coordination mechanisms to support sustainable environmental management.

This book, therefore, assembles case studies on institutional diversity from across the world to help conceptualize environmental institutions that are flexible enough to keep pace with societal and environmental changes. Our case studies exemplify global, national, and local level institutional diversity and illustrate how different forms of institutions interact historically and politically to influence the use of environmental resources. Each chapter identifies factors to consider before embarking on strategic institutional development in different socio-environmental contexts and provides examples of relevant policy instruments for regulating and monitoring environmental management. Some issues explored include inter-institutional gaps, path dependency, institutional isomorphism, bridging organizations, and institutional entrepreneurship. Lessons presented may be of value for an interdisciplinary audience working on and learning about complex environmental issues with a particular focus on social and economic equity and justice in environmental management.

Case studies are presented against the backdrop of interactional complexity to exemplify how complimentary, supplementary, and competing institutions can be categorized and characterized to support enhanced coordination. Drawing on the editors' previous work, this book proposes the use of cross-scalar, cross-cultural, and cross-functional dimensions to help conceptualize diverse institutional interactions. We argue that considering these dimensions may help to highlight different strategies for reshaping and developing robust institutions capable of mitigating undesired consequences from competing and conflicting interactions. Thus, rather than suggesting prescriptive institutional strategies, this book aims to provide practitioners with the tools necessary to better understand institutional diversity in environmental management.

Acknowledgements

The initial idea for this edited collection emerged while the Editors were graduate students in the Sustainable Futures Lab at McGill University. The lab group is led by Prof. Gordon Hickey, who has generously provided a forward to this book. We are indebted to his guidance and encouragement on this project and throughout our academic careers. Over the course of developing this book, we have been fortunate to share in many discussions on institutional diversity with a range of individuals, including:

- The co-contributors to this book. The authors of each chapter bring a diversity of experiences navigating the diverse institutions that shape sustainable environmental development. Together, the variety of insights provided in each chapter have allowed us to identify a range of complementary approaches for analyzing and understanding institutional diversity. We are especially grateful to each contributor for their patience as we navigated several major events and delays in completing this project.

- Our expert reviewers provided meaningful feedback and enhanced the quality of the final book. These individuals include Elsa Berthet, Hekia Bodwitch, Samantha Darling, Swapan Kumar Sarker, Kristen Lowitt, Archi Rastogi, Arlette Saint Ville, Jennifer Spence, and Wesley Tourangeau.

- Our academic colleagues at McGill University, Dalhousie University, and the University of Saskatchewan have generously engaged in critical discussion of the concepts underlying this edited volume.

- Contributors and participants at a hosted session on Institutional Diversity at the 2022 International Association for Society and Natural Resources Conference.

Acknowledgments are provided in each chapter to thank the individuals and funders who helped make this work possible. We would also like to extend our gratitude to Stephanie L. Aitken who kindly provided us with the cover image for this text.

Contents

1

Introduction

The Increasing Relevance of Assessing Institutional Diversity in Sustainable Environmental Management

H. M. Tuihedur Rahman[a],* and *Ashlee-Ann Pigford*[b]

Institutions for Sustainable Environmental Management

We live in a world requiring long-term solutions to profound and complex environmental challenges. Examples of challenges include the effects of anthropogenic climate change, the overexploitation of land and natural resources, rapid loss of biodiversity, high ecological footprints, ecological imbalances, high levels of environmental pollution, poor air and water quality, and the unequal distribution of environmental resources, as well as unsustainable infrastructure (Corson 1994; Dentinho 2011; Mirza 2006; Rahman et al. 2024; van den Bergh and Verbruggen 1999). Sustainable environmental management[1] needs to account for the intricacies of socio-ecological systems (Chapin et al. 2009) while also

---

[1] Sustainable development is a paradigm "…that meets the needs of the present without compromising the ability of future generations to meet their own needs" (Brundtland 1987). It emerged as a concept in global discourse in the 1970s and is now widely accepted as a key pillar of global development. It has been defined and refined through several key reports and declarations: the Club of Rome Report *The Limits of Growth* in 1972, the Declaration of the United Nations Conference on the Human Environment in 1972, the Brundtland Commission Report *Our Common Future* in 1987, and the Rio Declaration on Environment and Development in 1992.

[a] Department of Agricultural and Resource Economics, College of Agriculture and Bioresources, University of Saskatchewan, Room 3D34, Agriculture Building, 51 Campus Drive Saskatoon, SK S7N 5A8 Canada.

[b] Alumni of the Department of Natural Resource Sciences, McGill University, Macdonald Campus, 21,111 Lakeshore, Ste-Anne-de-Bellevue, Quebec, Canada H9X 3V9.
Email: ashlee-ann.pigford@mail.mcgill.ca

* Corresponding author: hm.rahman@mail.mcgill.ca; ucd269@mail.usask.ca

considering the finite nature of environmental resources and the just distribution of the resources across and within societies (Swilling 2019). With no "one size fits all" solution readily available, many national and international targets, plans, strategies, and agreements have been created to help address these global issues. However, reflection on the implementation of initiatives like the United Nations' Sustainable Development Goals[2] suggests the need for improved approaches and models that better consider interlinkages between plans and within broader systems (Allen et al. 2018). Therefore, this book seeks to help make sense of how we analyze the diverse and interconnected institutions (e.g., political, economic, and social) that shape environmental management efforts.

The ways humans interact with their natural environments are guided by social institutions formed through rules, equilibriums, power structures, and belief systems. Institutions can be codified and clearly delineated and/or norm-based and uncodified (Lauth 2015; North 1991). They differ from one society to another, and once established, they often act as intermediaries between people and resources, providing structures that direct management decisions (Cumming et al. 2020). In the age of globalization, society-specific institutions cannot operate in isolation, as social and environmental systems are nested, and their management has cumulative impacts on the global environment. If left unacknowledged, interactions between and across different types of institutions can lead to poor institutional coordination and entrenchment, which can perpetuate the mismanagement of ecological systems (Herrfahrdt-Pähle and Pahl-Wostl 2012). Therefore, efforts to redesign environmental management approaches will rely on an improved understanding of the drivers of cooperative and/or competitive interactions between and across different institutions (Cumming et al. 2020; Ostrom 2009; Po et al. 2019; Rahman et al. 2019).

By studying institutional strategies, norms, values, and rules, institutional analysis can offer insight into how institutions influence environmental management outcomes (Ostrom 1990). Existing approaches for analyzing environmental institutions tend to focus on understanding targeted management areas and often point to neo-liberal explanations of institutional change (Rahman et al. 2017). Further, no single approach has been able to offer comprehensive insight into how to deal with the high level of institutional diversity characterizing environmental management (Cumming et al. 2020). Thus, there is a need to invite new perspectives, evidence, and explanations for institutional diversity that better reflect the nested and collaborative nature of social and environmental systems (Baird et al. 2019; Becker and Ostrom 1995; Po et al. 2019; Rahman et al. 2017). In this book, we argue for creating complementary analytical approaches that borrow from several schools of institutional thought, each bringing its interpretation of how to understand diverse, socially linked institutional priorities, histories, roles, and functions. This edited volume presents eight illustrative case studies to enhance conversations on institutional diversity in the context of connected and complex global environmental management. In this chapter, key ideas and debates surrounding scholarship on

---

[2] Many of the United Nations' Sustainable Development Goals (SDGs) (2015–2030) are either directly (e.g., 92 indicators) or indirectly linked to environmental sustainability considerations (UN Environment Program 2021).

institutional diversity are introduced, and areas of intellectual overlap and disjuncture are highlighted before proposing considerations that may enhance the analysis of institutional diversity in environmental management.

Theoretical Approaches to Institutional Analysis

This section summarizes how three common theoretical approaches for institutional analysis conceptualize institutions and make sense of institutional diversity in environmental management.

Rational Choice Institutionalism

Rational choice analysis is the most common approach to institutional analysis and focuses on understanding the role of individual behavior in creating collective outcomes. It examines "…the consequence of the aggregation of the decisions of many rational individuals maximizing their egoistic interests" (Levi 2009). More specifically, it argues that rational individuals possess preferences over different social alternatives, have clear knowledge about the world around them, and can use the knowledge to pursue their own interests (Shepsle 1989). This theoretical approach postulates that institutions result from an *equilibrium* between the choices and behavior of rational actors that ultimately create exogenous *constraints* or *rules* for regulating individual behavior (Farrell 2018; Greif and Kingston 2011a; Shepsle 2006). Therefore, they are viewed as the socially constructed, mutually agreed upon sets of rules, acts, and policies that regulate human behavior (North 1991).

Rational choice institutions range from well-structured in terms of endurance, interactions, and objectives of actors to less structured, amorphous, and implicit institutions (Shepsle 2006). Most structured institutions tend to be formal and characterized by codified rules resulting from constitutional processes that are enforced by a hierarchically organized official system, distributed across different scales and levels (Greif and Kingston 2011b; Lauth 2015). Examples include parliamentary or judicial systems, regulatory institutions, and environmental conservation agencies. Less structured institutions tend to be based on social customs, norms, and values and often emerge through informal interactions that determine what action is socially acceptable, ethical, and desirable (Casson et al. 2010; Ostrom 1990). Ultimately, multiple and diverse formal and informal institutions interact, seeking to maintain institutional efficiency for producing desired outcomes in the face of political, cultural, market, and biophysical changes (Ostrom 1998; Ostrom and Basurto 2011).

In the context of environmental management, the tragedy of the commons argues that in the absence of formal guiding institutions, rational actors will adopt strategies that maximize personal benefit, resulting in the overexploitation and depletion of common environmental resources (Hardin 1968). More recent institutionalist thought has argued that the tragedy of the commons occurs in open-access resource systems with no rules for governing resource use behavior and can be avoided through collective action (Ostrom et al. 1994). Given that an individual cannot obtain perfect information in real-world situations, collective action institutions are beneficial in that they enable rational actors to develop, maintain, and enforce self-governing institutions based on past interactions, trust, and social norms that support conformity, even though these institutions may not optimize individual outcomes (Ostrom and Ahn 2008; Ostrom 2014).

Advancements in rational choice analysis have largely focused on developing frameworks to understand better the institutions governing socio-ecological systems across global, national, sub-national, and local scales. A set of design principles seeks to characterize the conditions under which collective actions generate durable environmental governance institutions (Becker and Ostrom 1995; Ostrom 1990). Further, Ostrom's Institutional Analysis and Design framework and Social-Ecological Systems framework help to illustrate how ecological, social, and governance properties interact to sustain and change resource systems (Ostrom et al. 1994). These frameworks identify complex systems of variables, rules, and external constraints and the polycentric nature of the institutional arrangements that influence resource governance (Ostrom 2011). Institutional polycentricity implies distributed decision-making autonomy across different levels (e.g., constitutional, collective-choice, and operational choice levels) (McGinnis 2011; Ostrom 2014). Rules developed at each level correspond with respective social, political, and environmental conditions, resulting in a wide variety of conflicting, cooperating, or overlapping rules. Understanding the diversity of rules and their association with social and environmental conditions is the central theme of rational choice institutionalism (Aligica 2014).

Despite having made important contributions to the understanding of institutions in environmental governance, the rational choice approach has struggled to analyze power relationships and account for historical interactions among diverse institutions (Clement 2010; Cumming et al. 2020; Hall 2009). For example, Rica et al. (2018) found that the rational choice approach could not explain national-local resource governing institution interactions, power relations among institutional actors, heterogeneity among resource users, and the socio-cultural and political contexts of a resource system. Since these limitations are inherent to rational choice theory, engaging with complementary theoretical approaches is important to capture a more comprehensive picture of the complex interactions between environmental systems and diverse institutions.

Historic Institutionalism

Historical institutionalism defines institutions as "formal or informal procedures, routines, norms and conventions embedded in the organizational structure of the polity or political economy" (Hall and Taylor 1996). One group of historical institutionalists knows institutions as long enduring and generally large-scale political structures in which actors are embedded, while the other group sees institutions as temporally changing processes like rules, norms, and strategies that regulate actors' behavior (Farrell 2018; Thelen 1999). The structural perspective suggests that social, political, and economic structures are context-specific and defined by a particular time and space. These structures organize society hierarchically, with individuals positioned through social networks, organizational structures, common practices, and shared meaning systems (Hall and Lamont 2013). Structural analysis aspires to understand how social, political, and economic power is distributed across institutional hierarchies and how the heterogeneous distribution of power can lead to a path of interactions among unequally empowered institutional actors (Pierson and Skocpol 2001; Scott 2008). The view of institution as a process aims to understand how institutions change as a response to unremarkable contingent events that bring about a significant alteration to an institutional structure over time (Thelen 1999; Mahoney 2000).

A historical approach to studying institutional change does not naturally intersect with rational choice analysis, which views institutions as rules or equilibriums that exist in response to exogenous shocks (Hall 2009). Instead, historical institutionalism identifies institutions as the outcome of evolutionary competitions among alternative forms of institutions and therefore supports causal explanations of institutional development and change (Pierson and Skocpol 2001; Greif and Kingston 2011). Three underlying propositions guide the temporal analysis of institutional development and change in a spatial context. First, it is believed that political events happen in a historical context, and earlier events influence later ones. Second, institutional actors learn about the behavior, strategies, and attitudes of their fellow actors over time. Third, experience can influence how actors predict the future and select interventions from many alternatives (Steinmo 2008). Ultimately, by revisiting historical interactions among institutional actors, historical institutionalists aim to understand the conditions determining why an action is taken and why a certain outcome emerges (Steinmo 2008).

Historical institutionalism offers a powerful conceptual and analytical approach to studying natural resources and environmental management (Holm 1995). Many studies present insight into how past management practices have obstructed the advancement of the types of institutional changes required to address underlying socio-environmental challenges related to poverty, social inequality, and environmental degradation (e.g., climate change, biodiversity loss, environmental pollution, etc.) (Rowbottom et al. 2022; Waylen et al. 2015). For instance, colonial institutional legacies characterized by hierarchical, non-participatory, and centralized institutional structures have been documented as key barriers to institutional change (Cockerill and Hagerman 2020; Mathur and Mulwafu 2018). Additionally, several studies indicate that resource use conflict is often rooted in the denial of multiple environmental management institutions with diverse historical origins, as well as a lack of collaboration among them (Dore 2001; Holst 2016). Further, historical intuitional analysis can help to explain how and why some institutional actors are empowered and endowed with legitimacy to make environmental decisions (Clement 2010), like in cases where Indigenous communities assume formal authority to govern and use natural resources (Alcantara et al. 2012).

The focus of historical institutionalism is on meso- and macro-level political processes operating within multiple institutional arrangements (Lieberman 2002). Rather than studying the overall institutional arrangement, competitive interactions among institutions representing different sets of "institutional logics"[3] are examined [for the explanation, see Thornton and Ocasio (2008)]. While interactions may

---

[3] Institutional logics are "…practices, beliefs and rules guiding an institutional order and providing actors with vocabularies of motive and sense of self" (Yu 2013).

move in any direction, stability is only achieved when interacting institutions follow a similar path; otherwise, institutions may change to generate new arrangements that better mediate or address conflicting interactions (Blyth et al. 2014; Mahoney 2000; Pierson and Skocpol 2001). As a standalone analytical approach, historic institutionalism can illustrate the emergence of institutional change over a temporal scale, but it does not offer sufficient insight into how institutional actors' interests shape the development of new institutions. Institutional analysts, thus, need to develop approaches that can help explain how diverse institutional actors learn and mobilize resources to navigate collaborative or conflicting interactions over time.

Organizational Institutionalism

Unlike the rational choice approach, organizational institutionalism (sometimes known as sociological institutionalism) does not assume that institutions are the products of purposive human action. Rather, institutions are gradually developed entities that provide a framework for deriving potential outcomes and are often taken for granted (DiMaggio and Powell 1983; March and Olsen 1998). Scott (2008) defines institutions as "… regulative, normative and cultural-cognitive elements that, together with associated activities and resources, provide stability and meaning to social life". These elements set standards for monitoring and sanctioning to ensure rule conformity by individuals, thus providing individuals with a decision-making framework for considering the consequence of making a choice to maximize personal benefits (Scott 2008; Wicks 2001). Organizational institutionalists examine how institutional logic, symbols, and myths guide collective framing for social organizations like corporate firms and local governments (Garud et al. 2007; Greenwood et al. 2014; Scott 2008). Greenwood et al. (2014) acknowledge the potential for multiple institutional logic, myths, and symbols to cause the emergence of multiple organizations, each composed of distinct institutional hierarchies. For instance, some institutions that oversee rule conformity may empower certain actors with authority to make judicial decisions, forming a new social hierarchy dissimilar to other social organizations operating within the same sphere (Scott 2008).

DiMaggio and Powell (1983) suggest that once diverse social organizations come to operate in a single area of interest, they may become similar through isomorphic processes. Isomorphic institutionalization, which mimics other institutions, can happen because of three underlying conditions (DiMaggio and Powell 1983). First, a supra-institutional structure that comprises multiple interdependent institutions imposes coercive authority and cultural expectations on an institution for legitimization. Second, institutions mimic each other to avoid ambiguity and symbolic uncertainty. The third source of isomorphic institutionalization is the normative professionalization of institutional actors to perform their roles. However, despite the creation of institutional isomorphisms through mimicry, scholars have argued for "institutional pluralism", suggesting that social actors and organizations operate under multiple autonomous institutions (Kratz and Block 2008; Thornton et al. 2012). This perspective posits that institutions do not always mimic each other to secure legitimacy; rather, legitimatization is a contested political process, posing new questions about how actors and organizations operate within multiple contested institutional logics (Yu 2013).

Environmental management scholars have used the concept of institutional isomorphism to identify and explain the context-specific impetus for adopting regulatory institutions for improving the environmental performances of social organizations (Kassim et al. 2020; Daddi et al. 2016; Rowe and Guthrie 2010). For instance, Daddi et al. (2016) observed that mimetic and normative drivers for adopting environmental regulations are more effective than coercive drivers among European organizations, such as businesses and social enterprises. In contrast, Kassim et al. (2020) found that coercive drivers positively influenced Malaysian local government organizations to adopt environmental management reporting and accounting. Organizational institutionalism can also explain how institutional innovation can act to dismantle isomorphic institutionalization, shedding light on how actors from diverse institutional identities leverage formal and informal networks to supplement and complement authority and redistribute roles, responsibilities, and resources to achieve common goals (Rahman et al. 2019). Since most environmental systems are multijurisdictional, their sustainable management requires the functional coordination and networking of diverse institutional actors, which can result in the innovative formation of new forms of institutions (Rahman et al. 2019). For example, actor interactions have led to the disruption and replacement of old institutions to help achieve collective commitments for reduced environmental pollution (Maguire and Hardy 2009), as well as the emergence of collective agreements to accommodate resources and to share roles and responsibilities for implementing innovative climate adaption options (Rahman et al. 2019).

Organizational institutionalism offers promising methodological and conceptual directions for studying organized actors (i.e., institutions as organizations). Yet, organizational institutionalism is limited to explaining how diverse institutions function in a single area of interest, focusing only on how social norms and values influence organizational behavior. Since individual environmental resource users are often organized by multiple and, at times, ambiguous social norms and values, hierarchical structure and norm-based interactions among individuals can be difficult to discern. Thus, it can be difficult to explain how both organized and unorganized social actors operate and participate in environmental management and shape institutional functions.

Opportunities for an Improved Understanding of Institutional Diversity

Analytical Framing: Three Dimensions of Institutional Diversity

This book argues for an enhanced approach to institutional analysis that accounts for multiple interpretations of diversity and embraces the complementary nature of existing theoretical approaches. Therefore, institutional diversity is considered a dynamic concept encompassing many types of multiples—across scales, levels, actors, and interpretations (Poteete 2012). Eight exploratory case studies illustrate factors to consider when designing institutional strategies for addressing environmental challenges. Collectively, the cases draw attention to the processes by which diverse institutions are built and reoriented through encounters with multiple perspectives, highlighting three common dimensions of institutional diversity: diversity across scales, cultures, and functions. This book posits that an improved understanding of these three dimensions may enhance the analysis of institutional diversity in environmental management. Before we present chapter summaries, we first identify several important opportunities for the study of institutional diversity related to each dimension:

- The cross-scalar dimension of institutional diversity is characterized by efforts to categorize and analyze environmental institutions from a variety of spatial, temporal, jurisdictional, and management scales (Cash et al. 2006; Gibson et al. 2000). Cross-scalar dynamics raise important questions associated with jurisdictional disjuncture, asymmetric power distribution, and the capacities of multiple institutions (Campbell and Hanich 2015; Po et al. 2019).

- The cross-cultural dimension of institutional diversity is characterized by culturally distinct environmental resource management institutions reflecting the unique language, knowledge systems, geographic distribution, and resource use practices of ethnically diverse communities (Maffi 2007). Considering cross-cultural aspects of institutional diversity can present insight into how diverse institutional interactions shape social equity, justice, and participation in environmental management.

- The cross-functional dimension of institutional diversity is characterized by either one institution performing diverse functions or multiple institutions performing the same function and is commonly observed among cross-sectoral institutions, where each institution is developed to serve sector-specific purposes (Young 1999). Cross-functional analysis can highlight institutional interactions, areas of ambiguity, and overlap, as well as the conditions necessary for institutional innovation.

Global Examples of Institutional Diversity: Chapter Contributions

By drawing on one or more of the theoretical approaches synthesized above, each chapter analyzes multiple and diverse institutions to enhance how institutional diversity is understood in diverse, polycentric institutional contexts. Table 1 summarizes the key elements of each case study, identifying the primary approach to institutional analysis taken. Each chapter adopts a unique approach to analyzing institutional diversity, considering different cross-scalar, cross-cultural, and cross-functional dimensions. While each chapter can be analyzed through the lens of multiple dimensions, this book seeks to highlight common insights. The book is organized by case study type, starting with three chapters that use the inter-institutional gap framework (IIG) (see Rahman et al. 2017), followed by three chapters that introduce complementary analytical approaches to enhance the analysis of institutional diversity (i.e., access theory, socio-cultural analysis, and robustness), before presenting two cases that consider institutions for global environmental governance.

Table 1. Summary of Case Studies.

Chapter # Authors	Case Study Focus/Location	Topic	Analytical Framing
2: Rahman et al.	Bangladesh	Indigenous land rights; cultural diversity, land management, national parks; inter-institutional gaps (IIGs); path dependency	Rational choice and historical institutionalism
3: Asuncion et al.	The Philippines	Mining for development; land ownership; political and civil rights; IIGs	Rational choice
4: Sarker et al.	Bangladesh	Tiger conservation; collaborative management; protected areas; IIGs	Rational choice
5: Bodwitch et al.	New Zealand	Commercial fisheries; Indigenous rights; co-governance; access theory	Rational choice
5: Saint Ville et al.	Jamaica	Food sovereignty; racial justice; Ital; socio-cultural perspectives	Historical institutionalism
7: Siddiki et al.	USA	Policy design; renewable energy; electricity; net metering; robustness	Rational choice
8: Rastogi et al.	Global	Multilateral institutions; climate finance; complexity	Historical and organizational institutionalism
9: Spence et al.	Arctic regional	Policy and decision making; sustainable development; environmental protection; bridging systems	Rational choice and organizational institutionalism

The first three case studies use the rational choice-based inter-institutional gaps (IIGs) framework (Rahman et al. 2017) to explore cross-scalar and cross-cultural dimensions of institutional diversity. In Chapter 2, Rahman, Pigford, and Natcher examine how institutions engage in path-dependent political interactions leading to ongoing land rights conflicts and forest destruction in Bangladesh's Madhupur Sal Forest (MSF). By comparing multi-scale formal policy instruments for land and forest management with local Garo community resource use and ownership practices, Chapter 2 reveals several inter-institutional gaps embedded in historical legacies that are contributing to present-day land disputes and forest destruction. Rational choice and historical path-dependency theories are combined to offer a conceptual and methodological approach that can systematically study the interactional gaps between government and Indigenous property rights institutions and help answer how the gaps evolve over time (Hall 2009; Mahoney 2000; Rahman et al. 2017).

In Chapter 3, Asuncion et al. present a case study on mining on Mindanao Island in the Philippines and identify the disjuncture between formal resource management rules, national government policies and actions, and Indigenous land management. International extractive industries supported by the national government have conducted mining activities in the ancestral lands of the Lumad community, who have organized strong resistance movements that have been met with state-sponsored militarization and violence. Using the IIG framework, the authors describe how divergent worldviews, a lack of negotiation among institutions distributed across scales, and failure to acknowledge and address cross-cultural institutional diversity can lead to socio-economic discrimination and marginalization.

Sarker et al., in Chapter 4, also adopt the IIG framework to study tiger conservation institutions of the Sundarbans Mangrove Forest (SMF), Bangladesh. Although the government of Bangladesh developed many policies and institutional practices for tiger conservation, habitat destruction and anthropogenic human-wildlife conflicts like poaching and retaliatory killing contribute to ongoing tiger population decline. The authors argue that when formal institutions (i.e., government policies) are authoritative and inadequately engage with local informal institutions rooted in cultural practices and value systems, there can be difficulty in arriving at a mutually agreed upon equilibrium. The IIG framework allows authors to illustrate issues with cross-cultural and cross-scalar interactions, suggesting that formal institutions would benefit from devising conservation actions that consider public protection and cultural practices.

In Chapter 5, Bodwitch et al. seek to address questions about access and institutional capacity as it relates to acknowledged Indigenous rights to use, own, and self-govern natural resources, more specifically, the allocation of commercial fishing rights in Aotearoa/New Zealand. The authors argue that improved coordination across institutional functions could address challenges related to equity, access, and ability to govern. By combining access theory with rational choice analysis, this chapter provides lessons for analyzing institutional diversity, pointing to the influence of cross-cultural, longstanding colonial legacies of exclusion.

In the sixth chapter, Saint Ville et al. examine the influence of the RastafarI socio-cultural movement in Jamaica on food sovereignty institutions. By adopting a historical institutional approach, the case study shows that RastafarI ideals spread region-wide and complement global food sovereignty discourse. This case highlights the interconnectivity between colonial legacies in an extractive economy and community self-determination, as well as important interlinkages between social institutions and environmental sustainability. Overall, this case study helps us better understand how diverse sociocultural institutions interact and influence each other over time.

Siddiki et al. provide a descriptive account of diverse institutional policies adopted by different jurisdictions to encourage the use of net metering in Chapter 7. The authors leverage ideas from rational choice analysis and institutional robustness to characterize the variation of policies adopted by different US states to encourage renewable energy production and consumption. This chapter contributes an expanded understanding of institutional analysis by considering institutional diversity, modularity, and redundancy.

The final two case studies focus on the organizational dimensions of modern international institutions. In Chapter 8, Rastogi et al. examine the historical and organizational development of multilateral and bilateral institutions designed to address climate change impacts and food insecurity. The authors contemplate underlying factors in creating isomorphic pathways and institutional redundancy, suggesting that rather than diversifying the purposive functions, isomorphic institutional development creates functional redundancy. In summary, Rastogi et al. illustrate that institutions must diversify their functions to remain relevant for serving emerging social and environmental needs.

In Chapter 9, Spence et al. adopt an analytical approach that combines institutional polycentricity from rational choice and organizational institutionalism to demonstrate how a single international organizational system can accommodate multiple functions and cultures to address diverse environmental challenges. The chapter focuses on three Arctic Council working groups: the Arctic Monitoring and Assessment Programme (AMAP), the Sustainable Development Working Group (SDWG), and the Emergency Preparedness, Prevention and Response Working Group (EPPR). The case study considers how actors from different cross-cultural institutional identities intersect to bridge different knowledge systems and communities of practice in support of environmental management in Arctic regions while illustrating how diversity can be accommodated in a single organizational structure by staying relevant and avoiding redundancy.

In the concluding chapter, Rahman and Pigford synthesize collective lessons from the case studies, arguing that an improved understanding of cross-scalar, cross-cultural, and cross-functional dimensions of institutional diversity may enhance the analysis of institutional diversity in environmental management. This chapter summarizes contributions to highlight important considerations for navigating diverse environmental challenges in polycentric institutional contexts. Several issues related to understanding the role of institutional diversity, inter-institutional gaps, path dependency, bridging organizations, and institutional entrepreneurship are covered in the context of environmental sustainability. The authors present examples of how to develop socially and politically relevant policy instruments for regulating and monitoring environmental resource use while considering local cultural and political processes as well as large-scale environmental management frameworks (national and global).

Acknowledgments

A-AP was funded through a Social Science and Humanities Research Council Post-Doctoral Fellowship at McGill University while this work was undertaken.

References

Alcantara, C., K. Cameron and S. Kennedy. 2012. Assessing devolution in the Canadian North: A Case Study of the Yukon Territory. Arctic 65(3): 328–338.

Aligica, P.D. 2014. Institutional diversity and political economy: the Ostroms and beyond. Oxford University Press: NY.

Allen, C., G. Metternicht and T. Wiedmann. 2018. Initial progress in implementing the Sustainable Development Goals (SDGs): A review of evidence from countries. Sustainability Science 13(5): 1453–1467.

Baird, J., R. Plummer, L. Schultz, D. Armitage and Ö. Bodin. 2019. How Does Socio-institutional Diversity Affect Collaborative Governance of Social-ecological Systems in Practice? Environ. Manag. 63: 200–214.

Becker, D.C. and E. Ostrom. 1995. Human Ecology and Resource Sustainability: The Importance of Institutional Diversity. Annual Review of Ecology and Systematics 26: 113–133.

Blyth, M., O. Helgadottir and W. Kring. 2014. Ideas and historical institutionalism. pp. 142–160. *In*: Fioretos, O. (ed.). Oxford Handbook of Historical Institutionalism. Oxford: Oxford University Press.

Brundtland, G.H. 1987. Our Common Future: Report of the 1987 World Commission on Environment and Development. Oslo: United Nations.

Campbell, B. and Q. Hanich. 2015. Principles and practice for the equitable governance of transboundary natural resources: cross-cutting lessons for marine fisheries management. Maritime Studies 14(1): 8.

Cash, D.W., W.N. Adger, F. Berkes, P. Garden, L. Lebel, P. Olsson et al. 2006. Scale and cross-scale dynamics: governance and information in a Multilevel World. Ecology and Society 11(2).

Casson, Mark C., Marina Della Giusta and Uma S. Kambhampati. 2010. Formal and Informal Institutions and Development. World Development 38(2): 137–141.

Chapin, III, F.S., G.P. Kofinas, C. Folke and M.C. Chapin 2009. Principles of ecosystem stewardship: resilience-based natural resource management in a changing world, Springer Science & Business Media.

Kassim, C.K., C.K. Hisam, S. Ahmad, N.E. Mohd Nasir, N.N. Mod Arifin and W.M.N.W. Mohd Nori. 2020. Environmental disclosures on local governments' websites: a malaysian context. International Journal of Public Sector Management 33(6/7): 663–679.

Clement, F. 2010. Analysing decentralised natural resource governance: proposition for a "Politicised" Institutional Analysis and Development Framework. Policy Sciences 43(2): 129–156.

Cockerill, K.A. and S.M. Hagerman. 2020. Historical insights for understanding the emergence of community-based conservation in Kenya: International Agendas, Colonial Legacies, and Contested Worldviews. Ecology and Society 25(2): art15.

Corson, Walter H. 1994. Changing Course: An Outline of Strategies for a Sustainable Future. Futures 26(2): 206–223.

Cumming, G.S., G. Epstein, J.M. Anderies, C.I. Apetrei, J. Baggio, Ö. Bodin et al. 2020. Advancing understanding of natural resource governance: a post-ostrom research agenda. Current Opinion in Environmental Sustainability 44 (June): 26–34.

Daddi, T., F. Testa, M. Frey and F. Iraldo. 2016. Exploring the link between institutional pressures and environmental management systems effectiveness: an empirical study. Journal of Environmental Management 183 (December): 647–656.

Dentinho, T.P. 2011. Unsustainable Cities, a Tragedy of Urban Infrastructure. Regional Science Policy & Practice 3(3): 231–247.

DiMaggio, P.J. and W.W. Powell. 1983. The iron cage revisited: institutional isomorphism and collective rationality in organizational fields. American Sociological Review 48(2): 147.

Dore, D. 2001. Transforming traditional institutions for sustainable natural resource management: history, narratives and evidence from Zimbabwe's communal areas. African Studies Quarterly 5(3): 1–18.

Farrell, H. 2018. The shared challenges of institutional theories: rational choice, historical institutionalism, and sociological institutionalism. pp. 23–44. *In:* Glückler, J., R. Suddaby and R. Lenz (eds.). Cham, Switzerland.

Garud, R., C. Hardy and S. Maguire. 2007. Institutional entrepreneurship as embedded agency: an introduction to the special issue. Organization Studies 28(7): 957–969.

Gibson, C.C., E. Ostrom and T.K. Ahn. 2000. The concept of scale and the human dimensions of global change: a survey. Ecological Economics 32(2): 217–239.

Greenwood, R., C.R. Hinings and D. Whetten. 2014. Rethinking institutions and organizations. Journal of Management Studies 51(7): 1206–1220.

Greif, A. and C. Kingston. 2011a. Institutions: rules or equilibria? pp. 13–43. *In:* Schofield, N. and G. Caballero (eds.). Political Economy of Institutions, Democracy and Voting, Berlin, Heidelberg: Springer Berlin Heidelberg.

Greif, A. and C. Kingston. 2011b. Institutions: rules or equilibria? pp. 13–43. *In:* Schofield, N. and G. Caballero (eds.). Political Economy of Institutions, Democracy and Voting, Berlin, Heidelberg: Springer Berlin Heidelberg.

Hall, P.A. 2009. Historical institutionalism in rationalist and sociological perspective. In Explaining Institutional Change: Ambiguity, Agency, and Power.

Hall, P.A. and M. Lamont. 2013. Why social relations matter for politics and successful societies. Annual Review of Political Science 16(1): 49–71.

Hall, P.A. and R.C.R. Taylor. 1996. Political science and the three new institutionalisms. Political Studies 44(5): 936–957.

Hardin, G. 1968. The Tragedy of the Commons. Science 162(3859): 1243–1248.

Herrfahrdt-Pähle, E. and C. Pahl-Wostl. 2012. Continuity and Change in Social-Ecological Systems: The Role of Institutional Resilience. Ecology and Society 17(2): art8.

Holm, P. 1995. The dynamics of institutionalization: transformation processes in Norwegian Fisheries. Administrative Science Quarterly 40(3): 398.

Holst, K. 2016. Colonial histories and decolonial dreams in the Ecuadorean Amazon. Latin American Perspectives 43(1): 200–220.

Kratz, M. and E. Block. 2008. Organizational implications of institutional pluralism. pp. 243–275. *In:* Greenwood, R., C. Oliver, K. Sahlin and R. Suddaby (eds.). The Sage Handbook of Organizational Institutionalism. Thousand Oaks, CA: Sage.

Lauth, H. 2015. Formal and Informal Institutions. pp. 56–69. *In:* Gandhi, J. and R. Ruiz-Rufino (eds.). Routledge Handbook of Comparative Political Institutions. Oxon: Routledge.

Levi, M. 2009. Reconsiderations of rational choice in comparative and historical analysis. pp. 117–133. *In:* Lichbach, M.I. and Alan. S. Zuckerman (eds.). Comparative Politics: Rationality, Culture, and Structure, Cambridge: Cambridge University Press.

Lieberman, R.C. 2002. Ideas, institutions, and political order: explaining political change. American Political Science Review 96(4): 697–712.

Maffi, L. 2007. Biocultural diversity and sustainability. pp. 267–277. *In:* Pretty, J., J. Guivant, T. Benton and A. Ball (eds.). The SAGE Handbook of Environment and Society, London: SAGE Publications Ltd.

Maguire, S. and C. Hardy. 2009. Discourse and Deinstitutionalization: The Decline of DDT. Academy of Management Journal 52(1): 148–178.

Mahoney, J. 2000. Path Dependence in Historical Sociology. Theory and Society 29(4): 507–548.

March, J.G. and J.P. Olsen. 1998. The institutional dynamics of international political orders. International Organization 52(4): 943–969.

Mathur, C. and W. Mulwafu. 2018. Colonialism and Its Legacies, as Reflected in Water, Incorporating a View from Malawi. WIREs Water 5(4).

McGinnis, M.D. 2011. An Introduction to IAD and the Language of the Ostrom Workshop: A Simple Guide to a Complex Framework. The Policy Studies Journal 39(1): 169–183.

Mirza, S. 2006. Durability and sustainability of infrastructure—a State-of-the-Art Report. Canadian Journal of Civil Engineering 33(6): 639–649.

North, D.C. 1991. Institutions. Journal of Economic Perspectives 5(1): 97–112.

Ostrom, E. 1990. Governing the Commons: The Evolution of Institutions for Collective Action. Cambridge: Cambridge University Press.

Ostrom, E. and T.K. Ahn. 2008. The meaning of social capital and its link to collective action. pp. 17–35. *In:* Svendsen, G.T. and S.L. Svendsen (eds.). Handbook of Social Capital: The Troika of Sociology, Political Science and Economics, Northampton, MA: Edward Elgar.

Ostrom, E., R. Gardner and J. Walker. 1994. Rules, Games, and Common-Pool Resources. Ann Arbor: University of Michigan Press.

Ostrom, E. 1998. A Behavioral Approach to the Rational Choice Theory of Collective Action: Presidential Address, American Political Science Association, 1997. American Political Science Review 92(1): 1–22.

Ostrom, E. 2009. A General Framework for Analyzing Sustainability of Social-Ecological Systems. Science.

Ostrom, E. 2011. Background on the institutional analysis and development framework. Policy Studies Journal 39(1): 7–27.

Ostrom, E. and X. Basurto. 2011. Crafting analytical tools to study institutional change. J. Inst. Econ. 7(3): 317–343.

Ostrom, V. 2014. Polycentricity: The structural basis of self-governing systems. pp. 45–60. *In:* Sabetti, F. and P.D. Aligica (eds.). Choice, Rules and Collective Action: The Ostroms on the Study of Institutions and Governance. ECPR Press, Colchester, UK.

Pierson, P. and T. Skocpol. 2001. Historical institutionalism in contemporary political science. pp. 693–721. *In:* Katznelson, I. and H.V. Milner (eds.). Political Science: State of the Discipline. Norton, New York.

Po, J.Y.T., A.S. Saint Ville, H.M.T. Rahman and G.M. Hickey. 2019. On institutional diversity and interplay in natural resource governance. Soc. Nat. Resour. 32(12): 1333–1343.

Poteete, A. 2012. Levels, Scales, Linkages, and Other "multiples" Affecting Natural Resources. IJC 6(2).

Rahman, H.M.T., S. Ingram and D. Natcher. 2024. The Cascading disaster risk of water, energy and food systems. Environmental Hazards, February, 1–20.

Rahman, H.M.T., A.S. Saint Ville, A.M. Song, J.Y.T. Po, E. Berthet, J.R. Brammer et al. 2017. A framework for analyzing institutional gaps in natural resource governance. International Journal of the Commons 11(2): 823–853.

Rahman, H.M.T., K. Sherren and van Proosdij. 2019. Institutional innovation for nature-based coastal adaptation: lessons from salt marsh restoration in Nova Scotia, Canada. Sustainability 11(23): 6735.

Rica, M., O. Petit and E. López-Gunn. 2018. Understanding groundwater governance through a social ecological system framework—relevance and limits. *In*: Villholth, K.G., E. López-Gunn, K.I. Conti, A. Garrido and J. van der Gun (eds.). Advances in Groundwater Governance. Leiden: CRC Press.

Rowbottom, J., M. Graversgaard, I. Wright, K. Dudman, S. Klages, C. Heidecke et al. 2022. Water governance diversity across europe: does legacy generate sticking points in implementing multi-level governance? Journal of Environmental Management 319(October): 115598.

Rowe, A.L. and J. Guthrie. 2010. The Chinese Government's formal institutional influence on corporate environmental management. Public Management Review 12(4): 511–529.

Scott, W.R. 2008. Institutions and Organizations: Ideas and Interests. 3rd ed. Thousands Oakes, California: Sage Publications.

Shepsle, K.A. 2006. Rational Choice Institutionalism. pp. 23–38. *In*: Rhodes, R.A.W., S.A. Binder and B.A. Rockman (eds.). The Oxford Handbook of Political Institutions. Oxford: Oxford University Press.

Shepsle, K.A. 1989. Studying institutions. Journal of Theoretical Politics 1(2): 131–147.

Steinmo, S. 2008. Historical institutionalism. pp. 113–138. *In*: Della, P.D. and M. Keating (eds.). Approaches and Methods in the Social Sciences. Cambridge: Cambridge University Press.

Swilling, Mark. 2019. The Age of Sustainability. London: Routledge.

Thelen, Kathleen. 1999. Historical institutionalism in comparative politics. Annual Review of Political Science.

Thornton, P.H. and W. Ocasio. 2008. Institutional logics. *In:* Greenwood, R., C. Oliver, R. Suddaby and K. Sahlin-Andersson [eds.]. The SAGE Handbook of Organizational Institutionalism. Sage Publications, London.

Thornton, P.H., W. Ocasio and M. Lounsbury. 2012. The Institutional Logics Perspective: A New Approach to Culture, Structure, and Process. Oxford University Press, Oxford, UK

UN Environment Program. 2021. Measuring Progress: Environment and the SDGs. Nairobi.

van den Bergh, J.C.J.M. and H. Verbruggen. 1999. Spatial sustainability, trade and indicators: an evaluation of the "Ecological Footprint". Ecological Economics 29(1): 61–72.

Waylen, K.A., K.L. Blackstock and K.L. Holstead. 2015. How does legacy create sticking points for environmental management? Insights from Challenges to Implementation of the Ecosystem Approach. Ecology and Society 20(2): art21.

Wicks, D. 2001. Institutionalized mindsets of invulnerability: differentiated institutional fields and the antecedents of organizational crisis. Organization Studies 22(4): 659–692.

Young, O.R. 1999. Governance in World Affairs. Cornell University Press, Ithaca and London.

Yu, K-H. 2013. Institutionalization in the context of institutional pluralism: politics as a generative process. Organization Studies 34(1): 105–131.

2

Path-Dependent Pathways of Inter-Institutional Gaps in Natural Resource Management

H. M. Tuihedur Rahman,[1,*] *Ashlee-Ann Pigford*[2] and
David Natcher[3]

Introduction

Institutions—the mutually agreed upon sets of rules that regulate human behavior (North 1991)—are the outcomes of cooperative and/or competitive interactions among actors who have a multiplicity of interests, perspectives, and agendas. Institutions are diverse and often conceptualized as formal and informal (Lauth 2015; North 1991; Rahman et al. 2017). Formal institutions are codified rules like laws, acts and policies emerging from international, national, or sub-national political processes. These rules are reinforced by well-structured and hierarchical bureaucratic systems, and deviation can trigger regulatory sanctions (Fukuyama 2013; Lauth 2015). Informal institutions are the outcomes of local political and cultural processes that develop into collective social memory, customs, and taboos. They are reinforced by cultural and behavioral practices held up by traditional and socially designated leadership and peer observation rather than organized bureaucracy

[1] Department of Agricultural and Resource Economics, College of Agriculture and Bioresources, University of Saskatchewan, Room 3D34, Agriculture Building, 51 Campus Drive Saskatoon, SK S7N 5A8 Canada.
[2] Alumni of the Department of Natural Resource Sciences McGill University, Macdonald Campus, 21,111 Lakeshore, Ste-Anne-de-Bellevue, Quebec, Canada H9X 3V9.
    Email: ashlee-ann.pigford@mail.mcgill.ca
[3] Department of Agricultural and Resource Economics College of Agriculture and Bioresources University of Saskatchewan, Agriculture Building, 51 Campus Drive, Saskatoon SK, S7N 5A8 Canada.
    Email: david.natcher@usask.ca
* Corresponding author: hm.rahman@mail.mcgill.ca; ucd269@mail.usask.ca

(Lauth 2015; Rahman et al. 2017). Notably, in some countries, informal institutions may be legally recognized within larger formal processes. For example, some Indigenous communities in Canada have signed self-government agreements with the federal government (Alcantara 2012; Natcher 2009). Both formal and informal institutions inform natural resource governance, each bringing resource management objectives and actions. When divergent formal and informal institutional objectives go unnoticed and unaccounted for over a long period, different social, political, and economic conflicts can emerge, leading to overexploitation and/or unjust use of natural resources (Helmke and Levitsky 2004; Rahman et al. 2017).

Historical legacies can be pivotal in shaping how natural resource management institutions undergo sequential, gradual, or transformative changes (Mahoney and Thelen 2010; North 1991). In this chapter, we consider how institutional legacies shape inter-institutional gaps (IIGs), which occur when there is an absence of mutually agreed-upon rules that could bridge the formal and informal institutional rules and practices (Rahman et al. 2017). Since IIGs are not static, they likely have undergone several modifications throughout history. We assume that if formal and informal institutions evolve independently, any unguided interactions between formal and informal institutions may compound and result in widening differences between the objectives and operations of different institutions. The evolution, nature, and legacy of such IIGs can be better understood by the historical analysis of formal and informal institutions. One approach to this type of analysis is to examine path dependency, which seeks to explain sequential change within institutions as a response to contingent historical events (Mahoney 2000).

This chapter seeks to enhance the analytical and explanatory capacity of the IIG framework (Rahman et al. 2017) by using a historical institutionalism approach to understand the origin of IIGs and their transformation over time. Path-dependency theory is leveraged to identify and characterize contingent historical events and their influences on inter-institutional interactions to help illustrate how institutions can become trapped in path-dependent, self-reinforcing and/or reactive pathways that may contribute to IIGs (Mahoney 2000). Theoretically, this chapter draws on both historical institutionalism (Hall 2009) and bounded rational choice-based institutional analysis (Ostrom 1998; Ostrom and Basurto 2011) to analyze a natural resource management case study from the Madhupur Sal Forest (MSF) of Bangladesh. The MSF is home to the Garo ethnic minority community, which is often in conflict with the government of Bangladesh over issues related to forest destruction. Therefore, the analytical approach adopted in this chapter may help explain how the conflicting interactions emerged and evolved and what broader social and environmental sustainability lessons we can draw from the case study. In the next two sections, we explain the theoretical foundation of the study, followed by a detailed historical description of the case study. Finally, we use the IIG framework and path-dependency theory to analyze the case study and highlight how IIGs evolve as a response to formal and informal institutional interactions.

Defining Inter-Institutional Gaps

The Inter-Institutional Gaps (IIGs) framework postulates that gaps emerge among natural resource management institutions in the absence of mutually agreed-upon rules (Rahman et al. 2017). Applying Ostrom's concept of institutional polycentricity to formal and informal institutions (Ostrom 2011), the IIG framework helps to reveal the political decision-making processes that drive disparity among socio-culturally distinct natural resource users (Rahman et al. 2017). The framework acknowledges that gaps surface when formal institutions fail to acknowledge and accommodate the resource management objectives of informal institutions and vice versa. These gaps can potentially be mediated when institutional actors coordinate objectives, craft knowledge networks, share power and authority, redistribute roles and responsibilities, and build mutual respect and reconciliation (Rahman et al. 2019). Overall, the framework helps identify where interventions can be taken to mediate the gaps.

The IIG framework draws on Ostrom et al.'s multi-level rule hierarchy described in the Institutional Analysis and Development (IAD) framework (Ostrom et al. 2016) and, subsequently, the Social Ecological Systems (SES) framework (Ostrom 2009). According to these frameworks, the institutions regulating resource manager and user activities are distributed into three layers of 'rules-in-use' (hereafter called rule levels): operational choice (i.e., day-to-day implementation of practical decisions by legally authorized actors, who are identified through a collective-choice process), collective-choice (i.e., process of making policy decisions by the authorized actors under the guidelines of the constitutional-choice process), and constitutional-choice (i.e., process of determining legitimate collective choices) rules (McGinnis and Ostrom 2014; Ostrom et al. 2016). Given that constitutional-choice level rules represent the overarching regulatory structure, Rahman et al. re-classified the rule levels into a binary category of constitutional and non-constitutional rules (encompassing collective-choice and operational-choice rules) based on their difficulty to differentiate and relative easiness to change with low transaction costs (Gordon et al. 2022; Rahman et al. 2017). Further, they identified four types of emergent gaps between the rule levels, each associated with a different underlying theory:

(a) A Gap Driven by Legal Pluralism: When formal constitutional rules fail to recognize existing informal constitutional rules and their actions, often resulting from differences between government-devised rules and local norms, values and customs regarding resource use (e.g., differences between local community traditions and national government regulations) (Rahman et al. 2014b).

(b) A Gap Driven by a Cultural Mismatch: When formal non-constitutional rules are formed with a lack of cultural understanding of informal constitutional rules, often created when the government's non-constitutional actors (e.g., local-level bureaucrats) fail to acknowledge local norms and customs or formal bureaucratic actors hold an imposition to control Indigenous lands and resources (Acheson 2006).

c) A Gap Driven by an Institutional Void: When formal constitutional rules deem the actions guided by informal non-constitutional rules as illegal or rules and policies are made the representation of all actors of interest. For instance, community leaders and other non-state actors (e.g., environmental activists, civil society organizations, intellectual communities, etc.) that come together outside of formal institutions to effect change (Hajer 2009) may seek to demonstrate their collective decisions to the government through either violent or nonviolent means (Hajer 2003).

d) A Gap Driven by a Structural Hole: When actions directed by formal non-constitutional rules diverge from actions guided by informal non-constitutional rules, often resulting from inadequate public participation in formal rule enforcement process or when formal institutions do not recognize local socio-political processes due to the lack of networking between formal and informal non-constitutional actors (Burt 2005).

The framework suggests that when there is a gap between formal and informal constitutional-choice rules (i.e., legal pluralism), it is likely that there is a gap between formal non-constitutional and informal constitutional-choice rules (i.e., cultural mismatch) (Rahman et al. 2017). Also, when there is a gap between formal and informal non-constitutional level rules (i.e., structural hole), there is likely to be another gap between formal constitutional and informal non-constitutional rules (e.g., institutional void) (Rahman et al. 2017). Such gaps are often expressed as social demonstrations and revolts leading to civil unrest (Hajer 2003). In order to illustrate the extent of gaps, the framework postulates that the gaps may be either co-existing,

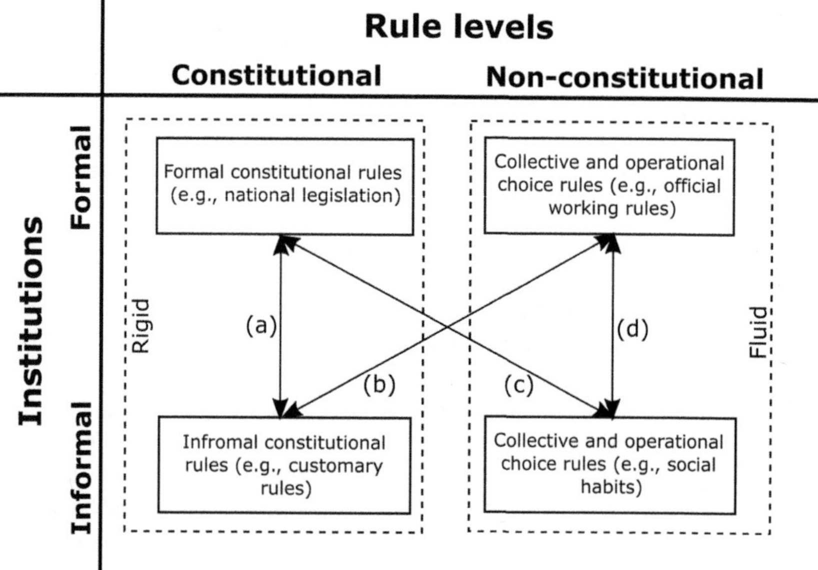

Figure 1: Inter-Institutional Gaps Framework (adapted from Rahman et al. 2017). (Note: a = legal pluralism; b = cultural mismatch; c = institutional void; d = structural hole).

latent or mediated. A gap is co-existing if there is no negotiation or interaction between the formal and informal institutions. Due to the presence of intermediary agents between formal and informal institutions to help build negotiation and cooperation, latent gaps remain obsolete until the intermediary agents disappear. Mediated gaps indicate that the formal and informal institutions are negotiated through mutually agreed upon rules for maintaining continuous communication, information exchange, knowledge and resource sharing, and the mutual agreement of co-governance of resource systems.

Path-Dependent Institutional Change

The IIG framework is based on rational-choice theory, which perceives institutions to be in a stable equilibrium among institutional actors involved in a bargaining process to agree on mutually acceptable actions (Farrell 2018). This approach explains institutional change as a response to exogenous shock leading to a new equilibrium but does not explain how new institutions emerge and overlooks the path-dependent nature of institutional change (Hall 2009). On the contrary, historical institutionalism seeks to understand how institutions change from one stable equilibrium into another by examining the brief and sporadic moment when a pre-existing institution collapses, opening an opportunity for change (Farrell 2018). Thus, the historical institutionalist approach has much to offer rational-choice approaches when investigating the development and modification of formal and informal institutions. In order to explain the process of institutional change, historical institutionalism emphasizes three key concepts: "critical juncture," "contingency" and "path-dependency" (Thelen 1999).

Critical Juncture

A critical juncture is a situation when the structural components of a political system are relaxed enough for political actors to have an array of possible alternative options to choose from for institutional change (Capoccia and Kelemen 2007). Identifying a critical juncture can only be done ex-post, implying that a critical juncture can be traced back as a causal foundation to progressive institutional change (Thelen 1999). Critical junctures appear for a short period (Capoccia and Kelemen 2007) because longer periods are often accompanied by structural constraints that limit options.

Contingency

During a critical juncture, political actors select one contingent option from several alternatives. The selection cannot predict its potential outcome either deterministically or probabilistically (Mahoney 2000), but once the choice is made, it sets a specific path-dependent trajectory for institutional development and change (Howlett and Rayner 2006). The selected option may not always reflect the relative advantage of the group of actors who bring about the change, although actors with potential advantage may be better positioned to influence the decision due to their agency, identity, and capacity (Hall 2009; Mahoney 2000). Thus, it is important to understand institutional structure at the critical juncture and the associated power distribution that shapes option selection to scrutinize the motivation for institutional change and identify the actors who may benefit from the change (Thelen 1999).

Path Dependency

A path-dependent trajectory is the chain of sequential events that result from contingent selections at a critical juncture. Path dependence can, therefore, be identified by "tracing a given outcome back to a particular set of historical events, and showing how these events are themselves contingent occurrences that cannot be explained on the basis of prior historical conditions" (Mahoney 2000). There are two types of path-dependent processes: self-reinforcing and reactive.

In a self-reinforcing sequence, initial change events (e.g., contingent selection of an option from alternatives) during a critical juncture initiate further changes in a direction that reinforces the new institutional structure to a level that becomes difficult to modify. Self-reinforcing path dependency is associated with decision "lock-in", which means a decision or action is believed to be better than other alternatives because enough people have already decided that the action or decision might result in increasing return (Page 2006). Once adopted, actors establish and maintain routinized activities they view as imperative for institutional legitimacy and stability (Mahoney 2000; Sydow et al. 2009), and those who benefit from the new arrangement will maintain its structure until they uncover a more attractive alternative. Therefore, although a "locked-in" institution might prove inefficient in changing conditions, its inherent self-reinforcing nature makes itself resistant to change.

A reactive sequence is a chain of causally connected and temporally ordered events (Mahoney 2000). Each event reacts to the preceding event and causes subsequent events, although the initial event at a critical juncture still follows the logic of contingency explained above. A reactive sequence differs from the self-reinforcing sequence in that the latter produces its institutional structure for solidification, while in a reactive sequence, a preceding event instigates a new event not to solidify itself but rather in response to a previous event. This response may revert the outcomes of earlier events, making it difficult to identify and characterize the initial events. In order to overcome analytical difficulty with reactive sequences, the concept of conjuncture is commonly used. A conjuncture is the intersection of two or more independent sequences of change (Mahoney 2000).

Depending on how a conjuncture modifies subsequent events, four potential scenarios can be assumed (Mahoney 2000):

(a) Two or more sequences of events may independently proceed without intersecting each other, and they may not inflict considerable institutional change.

(b) Two or more sequences of events for institutional change may intersect each other without giving considerable impetus for institutional change in any sequence.

(c) The independent sequences may intersect each other with significant impetus to change the causal trajectory of institutional change.

(d) If the conjuncture happens at a different time, the change may take a different trajectory than the one explained in the third scenario, and this temporal subjectivity of change forms the fourth scenario.

Path-Dependent Transformation of IIGs

Following the critical juncture and continent selection, the durability of a newly emerged institution depends on the level of conflict among the institutional actors during the juncture. Actors who fail to identify a mutually agreed upon change option may continue to oppose the change, resulting in a path-dependent trajectory of long-term institutional inefficiency (Capoccia 2016). Path dependency may be shaped by and create different types of IIGs based on how different levels of formal and informal institutions interact at the critical juncture. Given that the IIG framework is a relatively new institutional analysis tool, it does not capture what factors determine how informal institutional actors' respond to formal institutional actions during historical events. As such, it is also unknown how the gaps appear, to what extent and when.

We draw on the IIG framework and path-dependency theory to explore how the path-dependent interactions between formal and informal institutions may impact the extent of an IIG (Table 1). The extent of an IIG can be mediated when formal and informal institutions build coordination mechanisms during the critical juncture and selection period that promote convergent outcomes. In such a case, informal institutions may complement their formal counterpart to make rule enforcement

Table 1: Path Dependency of Institutional Interactions Influencing the Extent of IIGs.

Interaction Between Formal and Informal Institutions	Path Dependency	Extent of IIGs
Complimentary	Critical Juncture: Participation of both formal and informal institutions to make mutually agreed upon contingent choices Path Dependency: Self-reinforcement for convergent outcomes	Mediated: A gap is mediated when the formal institution is effective in building coordination with the informal institutions.
Accommodating or Competing	Critical Juncture: Participation of only powerful actors to make a contingent choice Path Dependency: Self-reinforcement for powerful actors and reactive path dependency for unempowered actors for divergent outcomes	Co-existing: A co-existing gap prevails when the formal institution is effective in rule development and enforcement, but the informal institutional actors oppose the core spirits of the formal institutional rules without directly violating them.
Substitutive	Critical Juncture: Absence of mediating agents Path Dependency: Reactive for the convergent outcome of both institutions	Latent: A gap may remain latent when an external agent takes up the role of broker to build communication between the formal and informal institutional rules and actors (Helmke and Levitsky 2004; Rahman et al. 2017).

effective (Helmke and Levitsky 2004; Rahman et al. 2017). However, the IIG mediation only exists if it brings mutually beneficial returns to both formal and informal institutions. Such mediations are observed when cultural, knowledge, socio-political, and socio-economic diversity are addressed, and all interested actors are involved in rule-making processes.

In contrast, IIGs coexist when formal institutions develop and enforce rules that informal institutional actors oppose but do not directly violate. In such a case, formal institutions may disregard informal institutions during the critical juncture period; thus, informal actors might have less influence in selecting decisions from contingent options. The formal actors might then follow a self-reinforcing path as it supports increasing returns for influential formal actors, or they may get locked into a patterned decision-making practice. Due to a lack of arbitration capacity, informal institutions may be accommodating, and formal institutions may exercise political authority even if it is economically discriminatory (Helmke and Levitsky 2004; Rahman et al. 2017). IIGs, to this extent, can also be seen when the formal institution is ineffective in rule enforcement, and informal institutional actors compete with the formal institutions by continuously violating formal rules and capitalizing on the formal institution's weaknesses. Thus, co-existing IIGs often prevail until a formal institution has an impetus to mediate with its informal counterpart.

IIGs may remain latent when an external agent appears and takes up the role of a broker/intermediary to build communication or negotiation between formal and informal institutional rules and actors (Helmke and Levitsky 2004; Rahman et al. 2017). However, intermediaries may not be present during a critical juncture period, and they may not influence contingent selection. Yet, they may appear when new socio-political or economic interests emerge and support informal institutions adopting substitutions that overcome formal institutions' ineffective rule enforcement. The appearance and survival of the external agent may be a reactive response to the formal institutions' ineffectiveness for a long period. However, if the mediating agent disappears, gaps may not remain latent (Helmke and Levitsky 2004; Rahman et al. 2017).

Case Study: The Madhupur Sal (Shorea Robusta) Forest (MSF)

This chapter draws on an empirical case study derived from managing the Madhupur Sal Forest (MSF) in Bangladesh. The MSF is located in central Bangladesh, is rich with biodiversity, and is made up of two major land types: higher elevation, forested *Chala* lands used for housing and horticulture, and lower elevation *Baid* lands used for paddy rice cultivation (Islam and Hyakumura 2019; Rahman et al. 2014a). The MSF forest, located adjacent to some of the most densely populated townships and villages in the country, has been subjected to tremendous anthropogenic pressure, resulting in rapid depletion of forest cover, illegal land encroachment, and land use conflicts (Iftekhar and Hoque 2005; Islam and Hyakumura 2019, 2021; Kabir et al. 2021; Rahman et al. 2014b). For instance, the MSF is home to the Indigenous Garo community, which is governed by its own norm and custom-based informal institutions that have been at odds with formal forest management regimes over time (Rahman et al. 2014b).

The Garo Community in the MSF

Garo is the largest plainland Indigenous community in Bangladesh and claims to be the earliest inhabitants of the MSF, having migrated from the Garo hills in India (Jalil and Oakkas 2012). Many traditional Garo cultural practices continue today and differ from the majority of the Bengali population. For example, Garo communities are matriarchal, and a newlywed bridegroom moves to the bride's house after marriage to become part of the bride's family. Joint family structures dominate, and nuclear families are seen rarely, although this is changing. Usually, the youngest daughter (i.e., *Nokna*) is the heiress of family property, and her husband is responsible for caring for family members (Jalil and Oakkas 2012). Nokna are responsible for managing household activities and earning a living (Dey et al. 2014), while males usually communicate with other social and external actors and hold other decision-making roles. This matriarchal system was the foundation for cultural leadership, land use practices and the distribution of resources among community members (Dey and Laila 2017).

A History of the Formal Management of MSF

Historically, the formal management of MSF can be categorized into three periods: precolonial, colonial, and postcolonial. The MSF was under the governance of the Mughal Empire during the precolonial period, and by the fall of the Empire, the whole management system was taken over by British colonial rules. The British Colonial rule ended in 1947, giving birth to two independent countries: India and Pakistan. Present-day Bangladesh—the then-eastern part of Pakistan, also known as East Pakistan—gained independence from Pakistan in 1971. Each period inherited rules from its predecessor, either in modified or in exact form, and had different implications for the Garo community. This section presents the historical evolution of MSF management rules.

Precolonial Mughal Period (1757)

The Bengal Subah region (present-day Bangladesh and the West Bengal province of India) belonged to the Mughal Empire and was ruled from the capital, Delhi, which was located approximately 1,300 km away (Misra 2018). Although the Mughals never had full control of Garo lands, revenues were collected by local kings and *Zamindars* (i.e., landlords) as the representatives of the emperor (Bal and Chambugong 2014; Cederlöf 2009; Misra 2018, 2021). During this time, Garo's traditional land rights and customary land management systems were left alone by external governments (Majid 2006). Garo were peasants and practiced Jhum (i.e., slash and burn) cultivation in the *Chala* lands of the MSF. They followed a family-based land management system where all members of a family would work and harvest agricultural fields before distributing the crop equally. Under this system, there are no private property or land selling or transfer rights (Rahman et al. 2014b). Garo belonged to maternal clans (i.e., *Mahari*) composed of a few families with blood relationships that lived in the same place, at times forming a village known as *Akhing*. The total land in an *Akhing* is *Azmali* property (i.e., common property), and a *Nokna* (i.e., head of a family) was allotted a piece of land by a *Mahari* leader for a single year (Khaleque 1992).

British Colonial Period (1757 – 1947)

The British East India Company—a group of British merchants—took control of the Bengal Subah in 1757 and sought to maximize revenue collection by imposing strict land management systems and appointing landlords (i.e., *Zamindars*) (Chakrabarti 1994; Islam et al. 2015). The Company enacted the Permanent Settlement Act of 1793 to promote revenue and agricultural intensification (Whitehead 2012; Wright 1954). While initially adopting the Mughal strategy of non-interference, the British exerted control over Garo lands following new interest in cotton and coal mines (Majid 2006; Misra 2018). Battles fought between 1820 and 1871 resulted in the annexation of Garo lands (Bal 2007; Cooper 1992; Kumar 2005), allocating the MSF to the Northeast Rangpur territory (Agnihotri 1994; Bal 2007). The Sepoy mutiny of 1857 ended the East India Company's rule, and the British Empire subsequently took formal control of the subcontinent (Peers 2011). The British government retained the *Zamindari* system but took several measures to formally institutionalize the management of Garo land under a uniform legal framework in the Bengal region. However, several important colonial acts influenced the historical evolution of MSF management and shaped the evolution of Garo society.

i. Indian Tenancy Act, 1878 and Forest Act, 1878

Government-backed religious conversion to Christianity and the Indian Tenancy Act (1878) promoted a shift from a communal land management system toward private land ownership (Rahman et al. 2014b). The colonial government made it legally mandatory to register land to facilitate revenue collection and allotted privatized land to Garo individuals who converted to Christianity (Majid 2006). Private Garo lands were first formally registered under the Indian Tenancy Act, which allowed Garo to register *Baid* and some *Chala* (Khan and Samadder 2012). However, this Act did not have any special provisions for ethnic minorities. The Forest Act was also enacted in 1878 to ensure government control of forest resources by declaring unoccupied forest lands as 'reserved forests' (Dey et al. 2014; Kumar 2005). This Act abolished customary forest uses and dismantled traditional ownership of forest resources (Kumar 2005). Both Acts dismissed the traditional *Azmali* property right (i.e., common property right), which resulted in significant changes to Garo land use and agricultural practices (Dey et al. 2014, Kumar 2005). For example, land privatization affected the authority of traditional leadership in land distribution and cultivation practices, as communal *Jhum* cultivation was discouraged.

ii. Bengal Tenancy Act, 1885

The Bengal Tenancy Act of 1885 recognized the customary land rights of forest-dwelling Indigenous communities, including Garo, stating that "Nothing in this act shall affect any custom, usage or customary right not inconsistent with, or not expressly or by necessary implication modified or abolished by, its provisions"

(Rashid and Satter 2006). This Act was further amended in 1928 to allow *Rayatys* (i.e., tenants) to transfer their land rights to a land buyer or a new occupant with the permission of Zamindars. Another amendment in 1938 gave *rayats* permanent ownership of the land (Raihan et al. 2009). This Act had two serious implications for MSF management. First, after losing the ability to fix land rent, the Zamindars increased the number of private land-holding *rayats*. Second, when the Zamindars recognized that the Zamindari system might be abolished after future independence, they allotted land parcels to some Garo community members only with verbal agreement, which increased the number of unregistered landowners (Farooque 1997; Rahman, et al. 2014b).

iii. Forest Act, 1927

The colonial government developed the Forest Act of 1927 to promote scientific forest management. Prior to this Act, forest lands were owned and managed by the *Zamindars*. Under the *Zamindari* system, the MSF was under the *Zamindari* control and was registered as a *debottor* property (i.e., a property gifted to God) (Khan and Samadder 2012). The leaders of different *Maharies* agreed with the *Zamindars* to cultivate and live in the MSF, and they were allowed to practice Jhum cultivation, given that they did not destroy the tree species (Khaleque 1992). The Forest Act authorized colonial forest officials to determine and designate a forested area as a "reserved forest" and prohibit access to and use of the forest by any individual without government permission. As a result, the Act brought most of the *Chala* land—where Garo households were located and traditional Jhum cultivation was practiced—under government control. As such, many Garo households and their traditional agricultural lands became illegal. Second, it pushed the Garo community to adopt a sedentary lifestyle in place of their traditional nomadic style (Dey et al. 2014; Rahman, et al. 2014b).

Postcolonial Period (1947-Present)

British rule ended in 1947, giving birth to two independent countries: India and Pakistan. The MSF fell in the province of East Pakistan, which later became independent Bangladesh in 1971. Both independence events considerably transformed the governance of the MSF as both regimes enacted several acts and took action based on respective political and policy contexts. Wet rice cultivation was also popularized among Garo to feed the growing population (Dey and Laila 2017) and was promoted by Christian Missionaries who offered Garo men free rice cultivation training, although Garo women traditionally managed most agricultural activities (Dey 2008). Thus, legal and political transformation after the independence of Pakistan and, subsequently, Bangladesh brought significant transformations in Garo agricultural land use practices and social institutions (Dey 2008; Dey et al. 2014; Dey and Laila 2017).

i. The East Bengal State Acquisition and Tenancy Act, 1950

The Pakistani government used this Act to abolish the *Zamindari* system in East Pakistan. *Rayats* in the MSF got permanent ownership of *Baid* lands (Rahman et al. 2014b), while forest lands (i.e., *Chala*) were transferred to the

Forest Department (FD). This Act dissolved customary Garo ownership of forest land and resources and fully restricted traditional *Jhum* cultivation (Dey and Laila 2017). So, Garo started cultivating horticultural species like pineapple, banana, etc., in the *Chala* lands, although using *Chala* land was illegal (Khaleque and Gold 1993). Wet rice cultivation was popularized during this time to feed the growing Garo population and to keep pace with local socio-economic and political changes (Dey and Laila 2017). Christian Missionaries played a major role in this agricultural transformation by offering free rice cultivation training to Garo men, although traditionally, Garo women managed most agricultural activities (Dey 2008). Thus, agricultural transformation contributed to a major shift in gender roles and power structures within Garo society (Dey 2008; Dey et al. 2014; Dey and Laila 2017). The Act also restricted land transfer rights specific to ethnic communities. Since Garo community members expected to have hereditary use rights on forest lands, the Act led to property rights disputes on *Chala (forest)* land (Khaleque 1992). This was further complicated in 1962 when the FD declared the MSF as a "reserved forest" and issued a gazette notification stating that all forest land claimants must submit their legal claims within six months. Most Garo failed to submit claims since the notification was not widely disseminated, and Garo community members had inadequate capacity to complete official formalities.

ii. The Bangladesh Wildlife (Preservation Amendment) Act of 1974

After independence in 1971, the Bangladesh government enacted the Bangladesh Wildlife Act 1974 with a specific provision for establishing national parks, wildlife sanctuaries and game reserves. This had implications for the MSF since the FD declared 8,436 ha of the MSF as the Madhupur National Park (MNP) in 1962 (Khaleque 1992; Rahman et al. 2014b). This Act, thus, gave legal support to the MNP (Islam and Hyakumura 2019) and prohibited all economic activities within the park boundary impacting Garo living within the park.

iii. Institutional Actions Taken During the Bangladesh Government Regime

Although Bangladesh ratified the Indigenous and Tribal Populations Convention (No. 107 of the International Labor Organization) in 1972, the government issued a formal eviction notice to Garo in the MNP in 1978, offering 1 acre of land and one thousand takas as compensation to every evicted Garo household (Khaleque 1992). However, the lands up for compensation had already been occupied by Bengali settlers who migrated from India, so the Garo community did not benefit and stayed in the MNP. A Tribal Welfare Association was established in 1978 to allow Garo to express their concerns regarding the government's eviction strategy; however, concerns were not addressed (Fernando 1997). In 1983, the government issued another eviction notice and notified residents that all land claims within the MNP must be submitted. Garo community members made land claims through the *Zamindar* system from the colonial period, but Bangladesh government officials denied most claims, stating that the documents were forgeries (Khaleque 1992).

   During the same period, the government undertook measures that caused loss in agriculture grazing and promoted non-timber forest product collection (e.g., rubber plantations and air force base construction). Then, in 1994, the Government of

Bangladesh amended the Bangladesh Forest Policy to include several declarations shaping MSF management (Rahman et al. 2014b). These included the government's (1) intention to expand rubber plantations in the forest area along with planting non-native fast-growing trees for forestry, (2) denial of Garo land rights and (3) a plan to expand ecotourism in the natural forest areas. Following this plan, the Forest Department of Bangladesh initiated a National Park Development project in 1999 and started building an eco-park within the MNP in 2003 (Dey et al. 2014; Rahman, et al. 2014b). This included a plan to build a boundary wall around the park, which would limit the access of Garo people to their agricultural lands, resulting in a mass demonstration. Police and park officials opened fire to control the demonstration, killing a Garo man (Rahman et al. 2014b). Without resolving land disputes, several development projects were subsequently initiated (e.g., the Integrated Protected Area Co-management Project and the Sustainable Forests and Livelihoods).

Since traditional Garo leadership could not negotiate and arbitrate land claims, a replacement form of leadership emerged. Young men with modern and Western education, known as "New Leaders", came to represent and organize Garo (Dey et al. 2014; Krishna 2011). Their role involved representing the community as well as managing local informal land management, including land right transfer, conflict mitigation and maintaining social order (Rahman et al. 2014b). While these leaders gained popularity among the Garo community, government officials often identified them as patrons of illegal land grabbing and forest destruction. Consequently, many encounter legal sanctions from the FD (Dey et al. 2014; Rahman et al. 2014b). Many Garo have converted to Christianity, and the church supported education, health, and agricultural development sectors in 2007; the local church established a voluntary legal and financial support wing in 2007 for Garo who frequently encounter legal sanctions from the FD (Rahman et al. 2014b).

Findings: Analysis and Discussion

In a previous study, the IIG framework was used to explain gaps between different levels of formal and informal institutions in the MSF (Rahman et al. 2017). While this work illustrated the contemporary land and forest management conflict between the government of Bangladesh and the Garo community, it lacked a comprehensive understanding of how the conflict and associated gaps emerged. Here, we use path dependency to help explain the historical nature of the IIGs. As such, we begin our analysis by identifying a key critical juncture that impacted both formal and informal institutions to see how the IIGs emerged and evolved.

Critical Juncture: Pakistani Independence

As described earlier, critical junctures are large and rapid and discontinue changes that impose a historically embedded causal effect on institutional change (Collier and Munck 2022). Decisions taken during a critical juncture from an array of contingent choices may directly influence subsequent outcomes (Mahoney 2021). Drawing on this presumption, we note that the end of British colonial rule and the subsequent independence of Pakistan was a key critical juncture that shaped extensive institutional revision in the MSF. We argue that the earlier colonial takeover does not

Table 2: Key Properties of a Critical Juncture as Reflected in MSF Institutions.

Properties of Critical Juncture (Mahoney 2021)	Reflections on MSF Management
Critical junctures are defined as contingent events.	The independence of Pakistan presented several potential paths forward and its subsequent legal and institutional changes.
Contingent events are of causal importance for a specific outcome.	The abolition of the Zamindari system and declaration of the MSF as a protected area that resulted from the independence of Pakistan shaped the conflicting interactions between the formal (i.e., government) and informal (i.e., Garo norms and values) institutions.
A contingent event is not expected to occur, but it does occur.	The Garo community could not anticipate how the independence of Pakistan and the abolition of the Zamindari system might affect their land claim.
Contingent choice may depend on agency, subjective discretion and deliberate choice of individuals or groups.	The overall land governance was the choice of national-level leaders who did not pay particular attention to ethnic and Indigenous land claims.
Outcomes are the results of gradual processes.	The IIGs at different levels of formal and informal institutions happened over the long course of postcolonial forest and land management.

qualify as a critical juncture since new formal rules only brought isomorphic changes to the imperial land governance system to facilitate revenue collection, and most colonial rules did not shift Indigenous land governance and land use practices like *Jhum* cultivation. To summarize how the independence of Pakistan transformed land governance represents a critical juncture in MSF management history, we consider five key properties of a critical juncture (Table 1).

First, the critical juncture reflects a choice whereby the Pakistani government chooses to abolish the *Zamindari* system when presented with alternative contingent options for land governance. Secondly, government decisions and actions resulted in conflicting interactions between the formal and informal land governing institutions, which formed the causal foundation of long-enduring land conflicts in the MSF (Soifer 2012). The abolition of the *Zamindari* system was a significant change to the official land governance system. Garo were unfamiliar with the new system, and government officials failed to develop official or unofficial channels of communication. Third, the Garo community experienced contingent events as exogenous shocks, which emanated from the outside, suddenly and unexpectedly reshuffling the elements and properties of their land use and governance system. This kind of contingent event may not be anticipated because prior expectations originate from pre-existing system functions or common-sense intuition (Mahoney 2021). For Garo community members, the government's changes for land resource management, like declaring the MSF as a protected area, were unexpected as they had not been consulted before the decision was made. Formal institutional changes in land resource management caused the government to appropriate forest land, denying traditional land claims and prohibiting the traditional *Jhum* cultivation system.

Fourth, contingent choices made by institutional actors reflected the agency, authority, power, subjective discretion, and deliberate actions of actors in advantageous positions (Mahoney 2021). In the case of the MSF, we observe that the national politicians who made the decision to declare the MSF as a protected area immediately after the abolition of the *Zamindari* system were not Garo and came mostly from the majority population. Since Garo leadership lacked adequate power, authority and representation to influence national political leadership, the notion of Garo matriarchy was interpreted merely as female land ownership despite the extensive influence of the system on Garo society. Finally, the impacts of the critical juncture were not fully observed during the initial phases of regime shift. Rather, the gaps between the formal and informal institutions (described in the next section) surfaced and widened over time because of subsequent government decisions and changes to Garo society and institutions. The next section discusses how the IIGs gradually emerged and maintained a path-dependent trajectory.

IIGs in the MSF

All four co-existing IIGs have been identified in the MSF (Rahman et al. 2017). This section aims to unpack the origins of each IIG to support the empirical and theoretical advancement of the IIG framework.

Gaps Between Formal and Informal Constitutional-Choice Rules

The abolition of the colonial government and subsequent removal of the *Zamindari* system—the critical juncture in our analysis—accompanied a new institutional arrangement that restructured the ownership of the MSF. The new top-down formal institutional arrangement lacked a constitutional obligation to consult with Garo, living the MSF when developing new formal rules. The government, as the custodian of the forest, established itself as the sole authority for forest management, creating a protected area, national park, eco-park and other development projects. A series of sequential formal rules kept empowering constitutional-level decision-makers, which gave them a relative power advantage, allowing them to ignore Garo institutions.

Many of the formal rules were designed for forest conservation and to assimilate the Garo community into mainstream society, yet they contributed to widening the gaps between the constitutional level rules of both formal and informal institutions. This happened because government actions did not consider the cultural distinctiveness of Garo society, and the formal rules failed to legitimize Garo society. For example, land privatization dismantled clan-based land ownership, ignoring existing matrilineal land ownership and hereditary practices. It contributed to the disintegration of Garo families and traditional social structures. Consequently, the formal and informal constitutional rules remained noncontiguous throughout history.

Gaps Between Formal Non-Constitutional and Informal Constitutional Rules

Our case study reflects the historical nature of a cultural mismatch. During the British colonial regime, the government did not significantly interfere with the Garo way of life and left the Zamindars to negotiate land management informally. However, postcolonial bureaucrats in both the Pakistan and Bangladesh governments sought land ownership and enacted a series of forceful actions to evict or relocate Garo people without understanding the cultural consequences of these actions. Government officials likely lacked historical and cultural knowledge about Garo society, and traditional Garo leadership likely lacked knowledge about the formal arbitration and negotiation processes, making interactions challenging. Garo leaders failed to comprehend the importance of official documentation of land registration during the colonial period. The lack of both formal and informal actors to organize and execute their respective responsibilities contributed to cultural mismatches between the formal government bureaucrats and the Garo communities, resulting in distrust.

Gaps Between Formal and Informal Non-Constitutional Level Rules

To implement government decisions, bureaucrats must build a collaborative network with local communities. In the case of the MSF, bureaucrats in both the Pakistan and Bangladesh governments had little legal capacity to solve land rights disputes and were employed as the enforcers of legal measures against the occupants of forest land, which has been preferred as a strategy to regulate land disputes.

The emergence of "New Leaders" representing Garo while negotiating with nationa-level political leaders and government bureaucrats can be viewed as a reactive response to ever-hardening government policies and legal systems. The role of new leadership has evolved over time, originally manifesting collective-choice rules, later providing dispute settlement within the community and dominating informal land markets. A few studies suggest that some new leaders have opportunistically used their social position to dominate Garo society and engage in forest destruction activities (Rahman et al. 2014b; Dey 2009). To regulate the influence and activities of the new leaders, the forest officials use legal instruments exacerbating mutual distrust, which creates gaps between formal and informal non-constitutional-choice rules.

Gaps Between Formal Constitutional and Informal Non-Constitutional-Choice Rules

Constitutional-level decision-makers in postcolonial governments lacked representation from the Garo community. They initiated interventions, but most of them failed as the formal institution has shown a lock-in attitude in terms of dealing with community responses toward government actions (e.g., maintaining the status quo of sole government authority on land-related decision making, eviction of Garo people and unsettling land disputes). The inability of government bureaucracy to deal with cultural diversity might also have contributed to this gap, as they could not inform the decision-makers regarding the Garo interests as they considered the Garo land claim illegal in terms of an existing legal framework.

In contrast, Garo's leadership lacked the capacity to engage in national-level decision-making and political processes at the beginning of the Pakistani regime. Despite their capacity to organize the community's collective choices, even the new leaders made negligible improvements to land conflicts. Thus, the interaction between formal constitutional and informal non-constitutional level rules has remained an ever-expanding gap.

IIGs and the Historical Interactions Between Formal and Informal Institutions

Discussion on the historical nature of the IIGs reveals that formal and informal institutions have been on two distinct pathways (Figure 2). After the critical juncture, the formal institution broke away from replicating the colonial regime but inherited many colonial rules. The postcolonial regime sought to develop formal institutions that treated every citizen in the country equally without considering cultural distinctiveness. This simplified commitment did not account for the complexity of the social and cultural identities that determine social choices. Here, formal institutional actors developed an institutional arrangement (i.e., a combination of rules and bureaucratic actors) that infused co-existing IIGs at every level of formal and informal institutions. As a reactive response to the series of government actions, the informal institution has gradually shifted its matrilineal spirit, traditional land use practice and social leadership. Our case study also reveals that the outcomes of formal and informal institutions are contingent on the effectiveness of rule enforcement. Rahman et al. (2014b) suggested that neither formal nor informal institutions are fully effective in enforcing their respective rules in the MSF.

Interactions between the formal and informal institutions have brought divergent outcomes to the government and the Garo community. The government, as a powerful institutional regime, has been successful in retaining its contingent choice of appropriating the sole authority and ownership of the MSF since the beginning of the postcolonial period (i.e., at a critical juncture). In contrast, the informal institution failed to secure its traditional structure, including land ownership, due to inadequate political representation during the period of critical juncture. It, therefore, can be said that the government, at the constitutional level, has been effective in accommodating informal constitutional rules. However, the enforcement of formal constitutional rules through formal non-constitutional level actors has not shown the same level of effectiveness. Due to the inadequate capacity of formal non-constitutional actors to engage with the Garo community, the informal non-constitutional actors have been maintaining a competing interaction (Helmke and Levitsky 2004; Mahoney and Thelen 2010). As such, the formal non-constitutional actors are directly enforcing formal constitutional rules, while the new leadership of informal institutions is organizing community members for organizing demonstrations and protests. Thus, our analysis suggests that the interaction between the formal and informal institutions in the MSF lies in an intermediate space between accommodating and competing interactions.

Uncertainties associated with landownership can lead to resource (over) exploitation and illegal land grabbing (Rahman et al. 2015). The New Leaders have been unable to abolish the authority of traditional leadership even though they organized local communities for a demonstration against the enforcement of formal rules that create distrust between government officials and the Garo community. The continuation of coercive rule enforcement has not given the informal institution an opportunity to stabilize. The evolution of the informal institution was contingent on the government's choices in terms of making decisions. For example, the community has seen several causally linked changes like modifying gender roles with the restriction

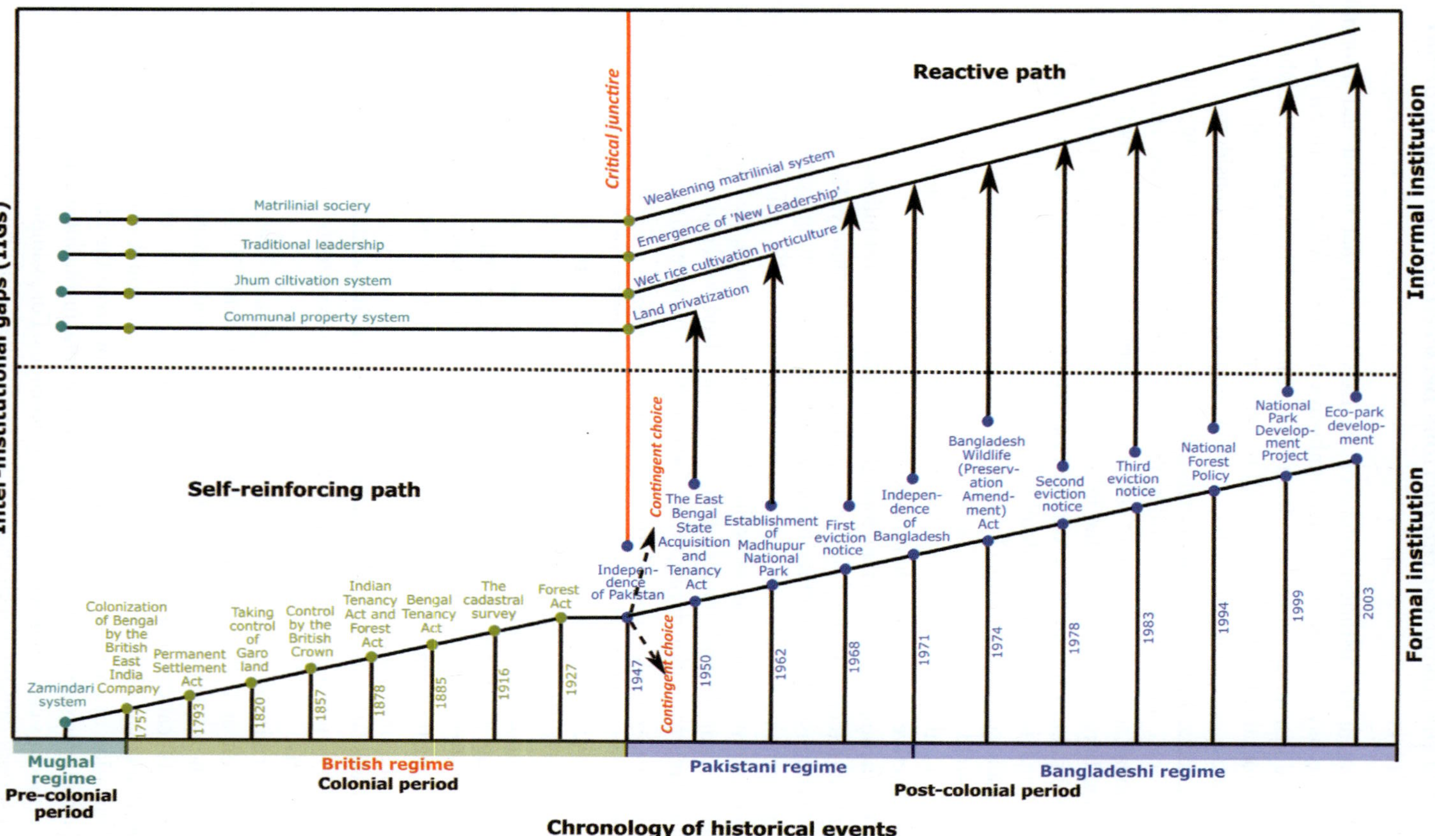

Figure 2: Historical Evolution of IIGs in the MSF.

of *Jhum* cultivation or the increasing political role of local churches in supporting local community members against FD legal sanctions. Also, the emergence of new leadership has added a layer of rules to the non-constitutional side of informal institutions when dealing with formal constitutional and non-constitutional rule-level actors. In many instances, such unsupervised interactions result in socio-economic discrimination and clientelism.

The development of authoritative rules throughout the postcolonial period can be viewed as evidence of a formal institution's self-reinforcement lock-in. Each formal institutional action supporting the lock-in appeared as an external force for informal institutional change. Informal institutions in the study area have been reacting to these forces by modifying their structural and operational arrangement rather than reproducing themselves (Figure 2). For example, the emergence of new leadership and changes in Garo's institutional structure were neither preemptive nor contingent. Rather, they have emerged as the reactive response of Garo society to government interventions. Based on this evidence, it can be suggested that if a coercive and authoritative formal institution enters into a self-reinforcing path, a relatively less effective and weaker informal institution may choose a reactive path. The co-existing IIGs in the MSF (Rahman et al. 2017) may have resulted from the respective self-reinforcing and reactive paths of formal and informal institutional change.

Conclusion

The IIG framework offers an approach to understanding discrepancies between formal and informal institutions. In many instances, these discrepancies are historically rooted and difficult to mediate when their legacies are not well understood. This chapter aims to enhance the analytical capacity of the IIG framework to address the historical legacies of IIGs using a historical institutionalism approach. Taking the land use conflict of the MSF of Bangladesh as a case study, the chapter offers three important theoretical and empirical lessons.

First, contingent choices made at critical junctures without considering the choices of culturally distinct and poorly empowered communities may lay the foundation of future conflict. Therefore, national-level political leaders, despite their advantageous position in power structure, need to ensure the representation of all culturally distinct communities in institutional development.

Second, rule development and enforcement are two distinct institutional activities. A coercive and authoritative institution may find it easier to accommodate the conflicting choices of culturally distinct societies. However, such choices impose difficulty on rule-enforcing actors like local-level bureaucrats who must communicate and deal with local people. If they fail to gain the trust of local people, local-level bureaucrats with poor decision-making autonomy follow a self-reinforcing path of enforcing coercive rules. Such actions never improve cooperation between the government and the community but contribute to widening IIGs.

Finally, when a formal institution follows a self-reinforcing path of institutional development and changes in a conflicted natural resource system, the informal institution follows a reactive path by changing its structure. This happens because formal institutional change appears to be an external drive of change for the informal institutions. The reactive changes of informal institutions may bring unintended consequences to the formal institutional objectives and the community itself.

Overall, this chapter introduces an approach to studying the historical evolution of IIGs and offers lessons regarding the factors that need to be considered in the process of institutional development in a culturally diverse natural resource management context.

Acknowledgments

A-AP was funded through a Social Science and Humanities Research Council Post-Doctoral Fellowship during the course of this work.

References

Acheson, J.M. 2006. Institutional failure in resource management. Annual Review of Anthropology 35: 117–134.

Agnihotri, S.K. 1994. District councils under sixth schedule. Journal of the Indian Law Institute 36(1(January-March)): 80–89.

Alcantara, C., K. Cameron and S. Kennedy. 2012. Assessing devolution in the canadian north: A Case Study of the Yukon Territory. Arctic 65(3): 328–338.

Bal, E. 2007. Becoming the Garos of Bangladesh: Policies of exclusion and the ethnicisation of a 'tribal minority'. South Asia: Journal of South Asia Studies 30(3): 439–455.

Bal, E. and T.C. Chambugong. 2014. The borders that divide, the Borders that Unite: (Re)interpreting Garo Processes of Identification in India and Bangladesh. Journal of Borderlands Studies 29(1): 95–109.

Burt, R.S. 2002. The social capital of structural holes. pp. 148–190. *In*: Guillén, M.F., R. Collins, P. England and M. Meyer [eds.]. The New Economic Sociology: Developments in an Emerging Field. Russell Sage Foundation, New York.

Capoccia, G. 2016. Critical juncture. pp. 89–106. *In*: Fioretos, K.O., T.G. Falleti and A.D. Sheingate (eds.). The Oxford Handbook of Historical Institutionalism. Oxford University Press, Oxford.

Capoccia, G. and R.D. Kelemen. 2007. The study of critical junctures: Theory, narrative, and counterfactuals in historical institutionalism. World Politics 59(3): 341–369.

Cederlöf, G. 2009. Fixed boundaries, fluid landscapes: British expansion into Northern East Bengal in the 1820s. Indian Economic and Social History Review 46(4): 513–540.

Chakrabarti, S. 1994. Collaboration and Resistance: Bengal Merchants and the English East India Company, 1757–1833. Studies in History 10(1): 105–129.

Collier, D. and G.L. Munck. 2022. Introduction: Tradition and innovation in critical juncture research. pp. 1–32. *In*: Collier, D. and G.L. Munck (eds.). Critical Junctures and Historical Legacies: Insights and Methods for comparative social science. Rowman and Littlefield, Maryland.

Cooper, J. 1992. The Garo of Bangladesh: A forest people's struggle to survive. Ethnic and Racial Studies 15(1): 85–101.

Dey, S. 2008. Deforestation and the Garo women of Modhupur Garh, Bangladesh. Asian Women 24(3): 57–81.

Dey, S and R. Laila. 2017. Reconstruction of women's role in jhum cultivation and the shift in the gendered division of labor among the Garos. pp. 8–39. *In*: Guhathakurta, M. and A. Banu (eds.). Gendered Lives, Livelihood and Transformation: The Bangladesh Context. The University Press Limited, Dhaka.

Dey, S., B.P. Resurreccion and P. Doneys. 2014. Gender and environmental struggles: voices from Adivasi Garo community in Bangladesh. Gender, Place and Culture 21(8): 945–962.

Farooque, M. 1997. Law and custom on forest of Bangladesh: issues and remedies. Bangladesh Environmental Lawyers Association, Dhaka, Bangladesh.

Farrell, H. 2018. The shared challenges of institutional theories: Rational Choice, Historical Institutionalism, and Sociological Institutionalism. pp. 23–44. *In*: Glückler, J., R. Suddaby and R. Lenz (eds.). Knowledge and Institutions. SpringerOpen, Cham, Switzerland.

Fernando, J. 1997. Sustainable Development: Unwitting Ally of Nationalism and Bane for Ethnic Minorities. Sugar: A South Asia Research Journal 4(2): 31–51.

Fukuyama, F. 2013. What is governance? Governance 26(3): 347–368.

Gordon, J.Y., A.H. Beaudreau, E.M. Saas and C. Carothers. 2022. Engaging formal and informal institutions for stewardship of rockfish fisheries in the Gulf of Alaska. Marine Policy 143: 105170.

Hajer, M. 2003. Policy without polity? Policy analysis and the institutional void. Policy Sciences 36(2): 175–195.

Hajer, M.A. 2009. Authoritative Governance: Policy Making in the Age of Mediatization. Oxford University Press, Oxford.

Hall, P.A. 2009. Historical Institutionalism in Rationalist and Sociological Perspective. pp. 204–223. *In:* Mahoney, J. and K. Thelen (eds.). Explaining Institutional Change: Ambiguity, Agency, and Power. Cambridge University Press, New York.

Helmke, G. and S. Levitsky. 2004. Informal institutions and comparative politics: a research agenda. Informal Institutions and Comparative Politics Problematic 2(4): 725–740.

Howlett, M. and J. Rayner. 2006. Understanding the historical turn in the policy sciences: A critique of stochastic, narrative, path dependency and process-sequencing models of policy-making over time. Policy Sciences 39(1): 1–18.

Iftekhar, M.S. and A.K.F. Hoque. 2005. Causes of forest encroachment: An analysis of Bangladesh. GeoJournal 62(1–2): 95–106.

Islam, K.K. and K. Hyakumura. 2019. Forestland concession, land rights, and livelihood changes of ethnic minorities: The case of the Madhupur sal forest, Bangladesh. Forests 10(3): 288.

Islam, K.K. and K. Hyakumura. 2021. The potential perils of Sal forests land grabbing in Bangladesh: an analysis of economic, social and ecological perspectives. Environment, Development and Sustainability 23(10): 15368–15390.

Islam, S., G. Moula and M. Islam. 2015. Land rights, land disputes and land administration in bangladesh—a critical study. Beijing Law Review 06(03): 193–198.

Jalil, M.A. and M.A. Oakkas. 2012. The family structure and cultural practices of Garo community in Bangladesh: An overview. Himalayan Journal of Sociology and Anthropology 5: 95–110.

Kabir, K.H., A. Knierim and A. Chowdhury. 2021. No forest, no dispute: the rights-based approach to creating an enabling environment for participatory forest management based on a case from Madhupur Sal Forest, Bangladesh. Journal of Environmental Planning and Management 64(1): 22–46.

Khaleque, K. and M.A. Gold. 1993. Pineapple agroforestry: An indigenous system among the Garo community of Bangladesh. Society and Natural Resources 6(1): 71–78.

Khaleque, T.M.K. 1992. People, forests and tenure: The process of land and tree tenure change among the Garo of Madhupur Garh forest, Bangladesh. PhD thesis, Michigan State University, Michigan, USA.

Khan, A. and M. Samadder. 2012. Weeping of the forest: Unheard Voices of Garo Adivasi in Bangladesh. International Journal on Minority and Group Rights 19(3): 267–290.

Krishna, A. 2011. Gaining access to public services and the democratic state in India: Institutions in the middle. Studies in Comparative International Development 46(1): 98–117.

Kumar, S. 2005. State 'Simplification': Garo Protest in Late 19th and Early 20th Century Assam. Economic and Political Weekly 2(27): 2941–2947.

Lauth, H.-J. 2015. Formal and informal institutions. pp. 56–69. *In:* Gandhi, J. and R. Ruiz-Rufino (eds.). Handbook of Comparative Political Institutions. Routledge, London, UK.

Mahoney, J. 2000. Path dependence in historical sociology. Theory and Society 29(4): 507–548.

Mahoney, J. 2021. The logic of social science. Princeton University Press, New Jersey.

Mahoney, J and K. Thelen. 2010. A theory of gradual institutional change. pp. 1–37. *In:* Mahoney, J. and K. Thelen (eds.). Explaining institutional change. Cambridge University Press, Cambridge.

Majid, M. 2006. Garo ethnicity (in Bengali). Mawla Brothers, Dhaka.

McGinnis, M.D. and E. Ostrom. 2014. Social-ecological system framework: Initial changes and continuing challenges. Ecology and Society 19(2): 30.

Misra, S. 2018. The sovereignty of political economy: The Garos in a pre-conquest and early conquest era. Indian Economic and Social History Review 55(3): 345–387.

Misra, S. 2021. Peasants, Colonialism, and Sovereignty: The Garo rebellions in eastern India. Modern Asian Studies 55(5): 1681–1717.

Natcher, D.C. and S. Davis. 2007. Rethinking devolution: challenges for aboriginal resource management in the yukon territory. Society and Natural Resources 20(3): 271–279.

North, D.C. 1991. Institutions. Journal of Economic Perspectives 5(1–Winter): 97–112.

Ostrom, E. 1998. A Behavioral Approach to the Rational Choice Theory of Collective Action: Presidential Address, American Political Science Association 1997. American Political Science Review 92(1): 1–22.

Ostrom, E. 2009. A general framework for analyzing sustainability of social-ecological systems. Science 325(5939): 419–422.

Ostrom, E. 2011. Background on the institutional analysis and development framework. Policy Studies Journal 39(1): 7–27.

Ostrom, E. and X. Basurto. 2011. Crafting analytical tools to study institutional change. Journal of Institutional Economics 7(3): 317–343.

Ostrom, E., R. Gardner and J. Walker. 1994. Rules, Games, and Common-Pool Resources. The University of Michigan Press, Ann Arbor.

Page, S.E. 2006. Path Dependence. Quarterly Journal of Political Science 1(1): 87–115.

Peers, D.M. 2011. Sepoy Mutiny (1857–1859). *In*: Martel, G. (ed.). The Encyclopedia of War. Blackwell Publishing Ltd., Oxford.

Rahman, H.M.T., J.C. Deb, G.M. Hickey and I. Kayes. 2014a. Contrasting the financial efficiency of agroforestry practices in buffer zone management of Madhupur National Park, Bangladesh. Journal of Forest Research 19(1): 12–21.

Rahman, H.M.T., S.K. Sarker, G.M. Hickey, M.M. Haque and N. Das. 2014b. Informal institutional responses to government interventions: Lessons from Madhupur National Park, Bangladesh. Environmental Management 54(5): 1175–1189.

Rahman, H.M.T., G.M. Hickey and S.K. Sarker. 2015. Examining the Role of Social Capital in Community Collective Action for Sustainable Wetland Fisheries in Bangladesh. Wetlands 35(3): 487–499.

Rahman, H.M.T., J.Y.T. Po, A.S. saint Ville, N.D. Brunet, S.M. Clare, S. Darling et al. 2019. Legitimacy of different knowledge types in natural resource governance and their functions in inter-institutional gaps. Society and Natural Resources 32(12): 1344–1363.

Rahman, H.M.T., A.S. saint Ville, A.M. Song, J.Y.T. Po, E. Berthet, J.R. Brammer et al. 2017. A framework for analyzing institutional gaps in natural resource governance. International Journal of the Commons 11(2): 823–853.

Raihan, S., S. Fatehin and I. Haque. 2009. Access to land and other natural resources by the rural poor: The Case of Bangladesh. CIRDAP, Dhaka, Bangladesh.

Rashid, M. and F. Satter. 2006. Indigenous people's land right: Bangladesh perspective. International conference on land, poverty, social justice and development. Institute of Social Studies, The Hague.

Soifer, H.D. 2012. The Causal Logic of Critical Junctures. Comparative Political Studies 45(12): 1572–1597.

Sydow, J., G. Schreyögg and J. Koch. 2009. Organizational path dependence: Opening the black box. Academy of Management Review 34(4): 689–709.

Thelen, K. 1999. Historical institutionalism in comparative politics. Annual Review of Political Science 2: 369–404.

Whitehead, J. 2012. John Locke, Accumulation by Dispossession and the Governance of Colonial India. Journal of Contemporary Asia 42(1): 1–21.

Wright, H.R.C. 1954. Some aspects of the permanent settlement in Bengal. The Economic History Review 7(2): 204–215.

3

The Philippines' Neoliberal Extractive Industry

Mining for Development, State Violence, and Inter-Institutional Gaps in Resource Governance

Angela M. Asuncion,[1,*] *Nicolas D. Brunet,*[2]
H. M. Tuihedur Rahman[3] *and Dominique Caouette*[4]

Introduction

The mineral wealth of the Philippines is among the highest in the world, with an estimated value of approximately USD $1 trillion (Camba 2016). As a result, the country's vast mineral deposits have been and continue to be tremendously sought after by international actors and national governments with the means required to exploit these resources (Simbulan 2016). The Philippine government established strong commitments to large-scale mining policies in the mid-1990s to support this process, prioritizing foreign direct investments as the nation's principal catalyst for

[1] School of Environmental Design and Rural Development, University of Guelph, 50 Stone Road East, Landscape Architecture Building Rm 114, Guelph, ON, Canada, N1G 2W1.

[2] School of Environmental Design and Rural Development University, of Guelph, 50 Stone Road East Landscape Architecture Building Rm 114, Guelph, ON, Canada N1G 2W1,519 824 1450 x54414.
Email: brunet@uoguelph.ca

[3] Department of Natural Resource Sciences, McGill University, Macdonald Campus, 21,111 Lakeshore, Ste-Anne-de-Bellevue, Quebec, Canada H9X 3V9,Canada.
Email: hm.rahman@mail.mcgill.ca

[4] Department of Political Science, Asian and Indo-Pacific Studies Chair, University of Montreal, Science Politique, C-4026, Pav. Lionel Groulx, 3150 Jean Brillant, QC, Canada H3T 1NB, 513 343 6111 ext. 40899.
Email: dominique.caouette@umontreal.ca

* Corresponding author: angela.asuncionn@gmail.com

growth (Holden 2014; Khee-Jin Tan 2006). Philippine national policies have been firmly rooted in the mining for development (M4D) paradigm that prioritizes wealth creation and financing development through foreign investment (Camba 2015; Holden and Jacobson 2007; Magno 2015).

The mining for development paradigm is rooted in neoliberal ideologies of resource development (Graulau 2008; Holden and Jacobson 2013) and is characterized by free trade policies, state de-regulation, and foreign direct investment (Holden and Jacobson 2007). M4D is commonly applied in low-income, mineral-rich nations (Broad and Fischer-Mackey 2016), where neoliberal policies are expected to generate capital to finance development (Magno 2015).

However, rather than catalyzing development, some argue that the push for M4D has perpetuated abusive practices and corrupt and unaccountable behaviors, holding few benefits for the vast majority of the Philippines population (Graulau 2008; Magno 2015). In particular, foreign investments in extractive industries in the Philippines have created relationships of dependency, with debt obligations and foreign ownership producing degrees of geo-political control over the archipelago (Camba 2015; Holden and Jacobson 2013; Pedro et al. 2017). Power asymmetries, human and environmental rights abuses, rising poverty, and negligent mining operations from M4D policies have also generated resistance movements (Espiritu 2017). Since the late 1970s, Indigenous communities, marginalized religious groups, and economically poor Filipinos have mobilized against neoliberal mining policies (Aytin 2015; Holden and Jacobson 2013). To protect mining corporations, the government's response to resistance has often been coercion and political violence through state-sponsored militarization (Kreuzer 2009).

Issues of conflict, resistance, and state militarization are common global themes found throughout post-colonial states hosting extractive industries (Meger and Sachseder 2020). The Philippines constitutes a noteworthy case to analyze these issues further as the dispossession of Indigenous lands for resource extraction continues despite decades worth of resistance (Global Witness 2020). This chapter explores gaps within natural resource governance at the nexus of governmental agencies and anti-mining Indigenous communities in the Philippines. In doing so, we aim to contribute to community agency and mining literature by focusing on a case study of state violence within Indigenous communities in Mindanao, particularly within communities resisting mining in their ancestral domains. Specifically, we provide a comprehensive analytical perspective of resource governance within the Duterte administration. We use Rahman et al.'s (2017) inter-institutional gaps (IIG) framework as a tool to identify barriers in post-colonial governance between formal and informal institutions, which have led to the undermining and abuse of political rights and civil liberties in Lumad territories. The chapter begins with a summary of the M4D paradigm at a global scale. We then situate our case study by outlining the history and implications of mining in the Philippines, followed by a review of the IIG framework. We end with an analysis of state violence toward anti-mining Lumad communities in Mindanao, using the IIG framework to highlight the latent gaps within the mining sector.

The History of Mining in the Philippines

Since Spain colonized the Philippines in 1521 and the United States succeeding colonization in 1898, the country has been subject to an unruly history of plunder undertaken first by mercantilist interests and later by capitalistic ventures linked to imperialist powers (Holden et al. 2011). The nation's endowment in natural resources has been sought after by industrialized capitalist countries for centuries and built around center-periphery relationships between Global North actors and the Philippines, maintained through economic dependency (Escobar 2012). Nowadays, strong post-colonial relationships between the U.S. and the archipelago, and now China, are embedded in expanding international markets, accessing foreign direct investment to periphery resources, and U.S. military intervention (Simbulan 2016). The Philippines gained independence from the United States in 1945, yet neo-colonial powers allied with national elites, international markets, and transnational corporations continue to reign today.

International markets and foreign corporations have successively exercised neo-colonial powers over the archipelago (Camba 2015). The implementation of the World Bank's structural adjustment programs (SAPs) in the late 1970s led to privatization in Philippine state industries and forced policy restructuring to ensure foreign direct investment (Holden et al. 2011). With debt servicing accounting for 8–10 percent of the Philippine gross domestic product, SAPs resulted in massive unemployment and decreases in domestic social spending (Camba 2015). This historical period was not only relevant economically but politically as well. Following the mutiny and massive mobilization of civil society, this era came alongside the downfall and exile of former dictator Ferdinand Marcos, remembered as the People's Power Revolution that brought Corazon Aquino into power.

In 1995, the national government implemented the Philippine Mining Act, which was forcefully promoted by the United Nations Development Program and World Bank policies (Ciencia 2011; Rovillos and Tauli-Corpuz 2012). By 1996, the number of foreign mining companies investing in the Philippines increased by 400 percent (Holden and Jacobson 2013; Lindon et al. 2014). Today, this Act continues to allow 100 percent foreign ownership of mining operations, including various incentives to attract mining prospection (Aytin 2015; Magno 2015). The current administration (2016–2022) remains strongly committed to neoliberal mining policies, which is evident in President Duterte's April 2021 amendment to Executive Order No. 79. This amendment lifted the nine-year mining moratorium under the previous administration and enabled the government to enter new mineral agreements under the 1995 Mining Act (Ocampo et al. 2021).

For over two decades, the state's ongoing prioritization of foreign investment and rushed environmental assessments have led to disastrous socio-environmental consequences. The demand for accountability in the mining sector has been particularly strong among poverty-stricken populations and directly affected communities where M4D policies have not delivered expected economic returns (Simbulan 2016). From 1990 to 2005, the growth of gross domestic product per capita was at a mere 1.6 percent per year (Magno 2015). As a result of slow economic growth, the Philippines entered a spiral of heavy indebtedness, with the total value of the nation's external debt equal to 77 percent of all export revenue (Holden and Jacobson 2013). External debt has greatly hindered the achievement of poverty alleviation and increased social inequities. Poverty incidence in the Philippines continues to stand at 25–26 percent at the national level and 30–60 percent in large-scale mining provinces (Magno 2015).

This has been particularly true for culturally marginalized and underserved communities living in remote areas directly affected by mining operations (Figure 1). Unsurprisingly, many of these communities have been at the forefront of mining accountability campaigns, especially on the island of Mindanao. Lumad, translated from the Cebuano language, "born from the earth" refers to Indigenous peoples of the archipelago (Paredes 1997). The Lumad peoples of Mindanao are made up of approximately 18 distinct tribes (Rodil 2020). Lumad ethnic groups have traced the movements of generations in relation to the river systems they have occupied, with "social proximities to key bodies of water equated to civility, cultural purity, and political legitimacy" (Paredes 2016).

Although each of the 18 tribes is a discrete ethnolinguistic group from one another, Paredes (2016) describes the collective significance of water for Lumad peoples. Beyond a vital resource, water plays a significant role in the identity, oral tradition, ritual practices, and connection of socio-economic networks and power between tribes. The resource use practices of these communities are embedded in forest-based subsistence, such as hunting and gathering, swidden farming, and the collection of non-timber forest products, such as rattan and tree resin, for lowland selling (Paredes 2019).

With over 60 percent of mines operating in ancestral territories (Simbulan 2016), Lumad communities have been severely impacted by the externalities of mining and continue to be victims of development aggression, forced displacement, and extrajudicial killings (Global Witness 2020). A number of disasters and human rights abuses have led to strong Indigenous resistance movements (Espiritu 2017; Hagen and Minter 2019; Holden 2014). Government responses to such resistance movements have been met with state-sponsored militarization and violence (Caouette 2012; Iglesias 2018; Karapatan 2014). In several instances, public resistance to M4D policies has led to extrajudicial killings of Indigenous leaders and the classification of a number of Indigenous organizations and activists as terrorists and communist rebels (Simbulan 2016).

It is important to underline the current context of the increasing criminalization of Indigenous peoples and activists in the Philippines. Although specific to the Philippine context, this phenomenon reflects a global trend of state protection and support of private interests through coercion and violence. A 2020 Global Witness report recognized the Philippines as the second deadliest place in the world to be a land defender, with the mining industry being the most dangerous sector. The report further indicated that 43 land-rights advocates in the Philippines were killed in 2019, with most murders located in Mindanao. During the same period, according to government records, three mining applications for silver, gold, and copper projects were granted in the Mindanao region, which spans over 20,000 hectares (Global Witness 2020).

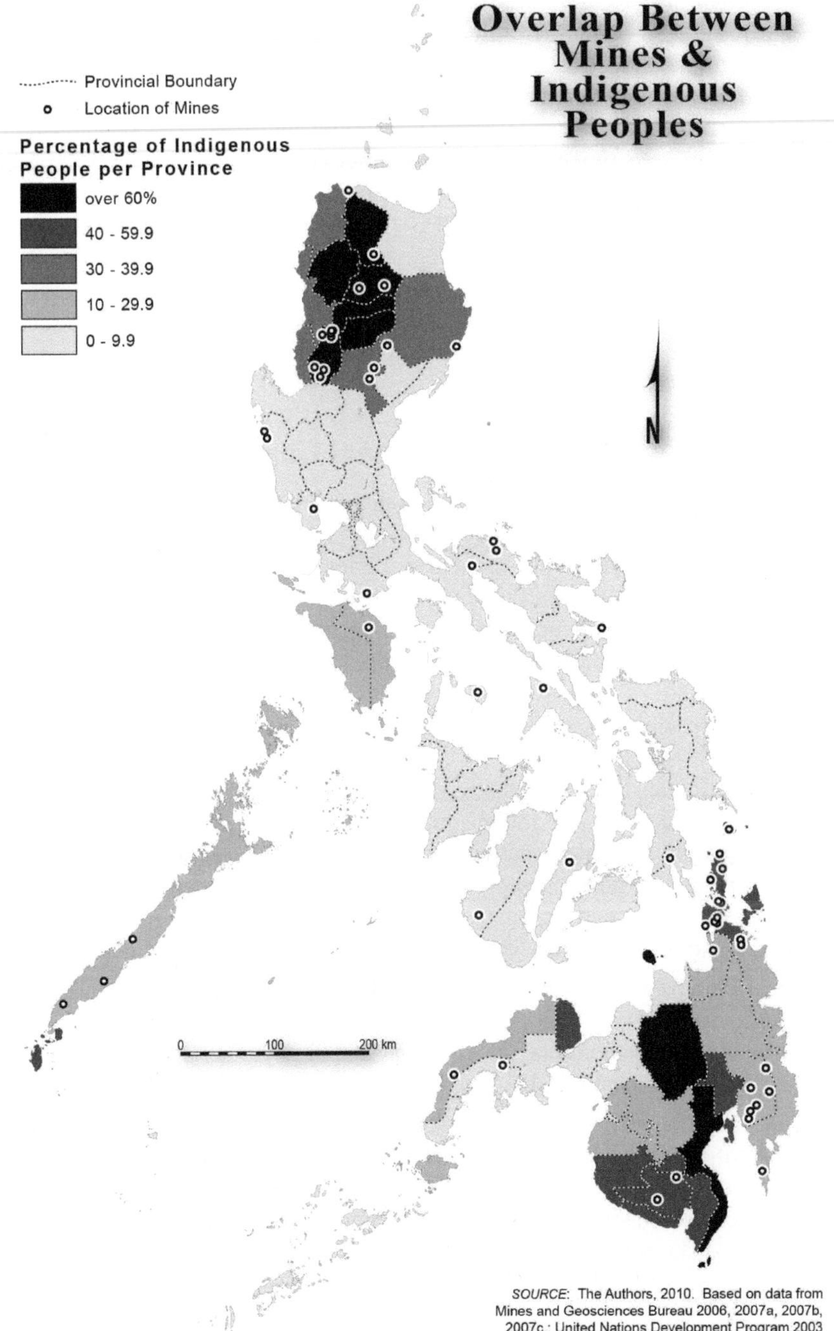

Figure 1: The overlap between mines and Indigenous peoples in the Philippines. Reprinted with permission from William Holden et al. (2011). Exemplifying accumulation by dispossession: mining and Indigenous peoples in the Philippines. Geografiska Annaler: Series B, Human Geography 93(2): 141–161.

In several instances, mining corporations request government military units or hire private armed groups to safeguard extractive operations and protect private interests in the face of resistance. The Armed Forces of the Philippines have been considered prone to and often dedicated to protecting foreign private investors from rebels, counterinsurgencies, and terrorist threats (Wilson 2019). The Special Civilian Armed Auxiliaries are trained by the Armed Forces of the Philippines but are usually paid by mining firms to protect projects (Holden and Jacobson 2007). The Citizen Armed Forced Geographical Units (CAFGU) is described by Zech and Eastin (2021) as a "type of pro-government militia that provides auxiliary support to the Armed Forces of the Philippines in waging counterinsurgency and providing community security in conflict zones." De facto, state-sponsored militaries are acting as security guards for private enterprises, posing violent threats to local activists and journalists alike (Kreuzer 2009).

Brief Overview of Inter-Institutional Gap Framework

This chapter uses the IIG framework (Rahman et al. 2019) to examine inter-institutional interactions between resource users and governing actors. Rahman et al.'s (2017) inter-institutional gap (IIG) framework is an analytical tool used to identify a lack of harmonization and conflict in common pool resource management. The framework's strength lies in its ability to analyze cross-cultural multi-level governance systems, which other frameworks have not yet been able to address (Rahman et al. 2017). Moreover, it emphasizes that long-term sustainable resource use can only be achieved through effective collaboration in multi-level governance systems.

The IIG framework's systematic matrix is used to analyze institutional gaps in resource governance. The framework uses institutional concepts, such as legal pluralism, cultural mismatch, institutional void, and structural holes, to account for the existing gaps in governance within socio-ecological systems. Drawing from Rahman et al. (2017) and Rahman et al. (2019), these four institutional concepts identified in the IIG framework can be described in the following manner:

a. Legal Pluralism: An inter-institutional gap emerges when formal constitutional rules, usually government-centric rules, do not acknowledge or value informal non-informal constitutional rules, actions, and/or traditions (e.g., a clash between national government legislation and Indigenous customary law).

b. Cultural Mismatch: This inter-institutional gap emerges when formal non-constitutional rules are created without regard to cultural values, customs, and norms of informal constitutional rules (e.g., local politicians and bureaucrats make choices based strictly on government rules without recognizing local established customs).

c. Institutional Void: A third type of inter-institutional gap emerges when formal constitutional rules treat the actions undertaken by informal non-constitutional rules as unlawful and punishable (e.g., activism and free speech outside of formal institutions are deemed illegal by government legislation and policies).

d. Structural Hole: This type of inter-institutional gap emerges when deviations occur between the actions implemented by formal non-constitutional rules and those advised by informal non-constitutional rules (e.g., a lack of public participation in formal non-constitutional decision-making).

Applying the IIG Framework to the Philippines

In the following pages, we explore how state violence and resource exploitation manifest as indicators of latent inter-institutional gaps in the Philippines' mining industry.

Methods

Peer-reviewed articles, national government mining legislation, news sources, and gray literature were gathered to inform our analysis of the Philippines mining sector (referred to as "the case"). Then, document analysis organized around key questions surrounding resource governance was undertaken to develop a broad understanding of the formal and informal institutions governing the nation's mining industry.

Identifying the Institutions Governing Mineral Resources in the Philippines

This section identifies the institutions and governing rules operating within the four spheres of the IIG framework, summarized in Table 1 (see below). In addition, we explore the formal and informal tools and mechanisms that actors possess to operationalize their role within the Philippines' mining industry.

Today, numerous diverse actors are involved in the Philippines' mining industry at the local, regional, national, and international levels. In an effort to examine the complex governance occurring in today's mining industry in Mindanao, we focused our attention on exploring the interactions between three groups of actors. These three main actors are central to understanding how mining extraction activities occur on the island. Specifically, we will be analyzing the interactions between the Duterte government, anti-mining Lumad communities, and the large-scale mining industries. The first two are that the government and the community organize mining systems through formal and informal institutions that possess de jure and/or de facto authority and legitimacy to enact rules. The third factor is the mining industry, which extracts mineral resources and operates under formal institutional regulations. The industry has an influential role to play in mining system governance.

In conflicting interactions, states and corporations gain momentum as community members are limited in their representation in local decision-making despite the Philippines' decentralized form of government put in place following the downfall of Marcos. Acheson (2006) describes this as the "concentration of the resource in the hands of local elites or corporations," an enabling feature common in post-colonial government policies. Government actors, many of whom are allied or friendly to mining corporations, therefore possess a relative advantage over informal institutional actors in regulating the mining system.

Table 1: Summary of the individual governing rules and institutions involved within the case's IIG framework analysis during the Duterte administration.

Governing Rules	Institutions	Summary	Main Rule Makers and Actors
Formal constitutional choice rules	Government's legislative and executive bodies	Philippine national government is characterized by a centralized weak state made up of large elite family conglomerates and led by a strong anti-communist socio-political environment	House of Representatives, Senate of the Philippines, President, Supreme Court, Department of Energy and Natural Resources (DENR), National Commission of Indigenous Peoples (NCIP), and local government units
Formal non-constitutional choice rules	State-sponsored militarization	The Philippine government created state-sponsored militarization units as defense mechanisms for those rebelling against the private interests of the ruling state	Armed Forces of the Philippines Citizen Armed Force Geographical Units (CAFGUs), Philippine National Police, and private operatives
Informal constitutional rules	Indigenous customary law	The Lumad peoples of Mindanao have been among the most impacted by the mining sector due to the location of ancestral domains. Customary law is embedded in sacred relationships with the natural world, with laws transmitted generationally through oral practice and traditions	Although there is no consensus on the complete list, Rodil (2020) notes that the Lumad Peoples in Mindanao are made up of about 18 tribes, including: - Ata - Bagobo - Blaan - Bukidnon o Talaandig - Dulangan - Isamal - Mandaya - Manobo - Banwaon - Mamanwa - Mandaya, Mansaka, Mangguangan and Dibabawon - Mansaka - Subanĕn - Tagakaolo - Tïboli - Tĕduray - Higaunon - Ubo
Informal non-constitutional rules	Indigenous livelihoods and cultural activities	Paredes (2016) states that the identities, networks, and livelihoods of Lumad people are embedded in river systems and understandings that the spirit of creation is within all aspects of the natural world	Elders, Tribal Chieftain, translated into *Datu (Male) Bae (Female)*, Indigenous community members (Socio-political structure of Lumad communities varies between tribes)

Inter-Institutional Gaps — Identifying Institutional Concepts

By examining the relationships between the aforementioned institutions, this section uses the IIG as an analytical framework (Figure 1) to examine why governance gaps exist and how they have manifested in increasing state violence and infringements upon Lumad civil liberties. This IIG framework highlights the interconnected gaps in governance between formal and informal institutions in the Philippines' extractive industry. Gaps are functions of the four possible institutional dynamics that can co-exist in this framework, with gap (a) as legal pluralism, gap (b) as cultural mismatch, gap (c) as structural hole, and gap (d) as institutional void (Figure 2).

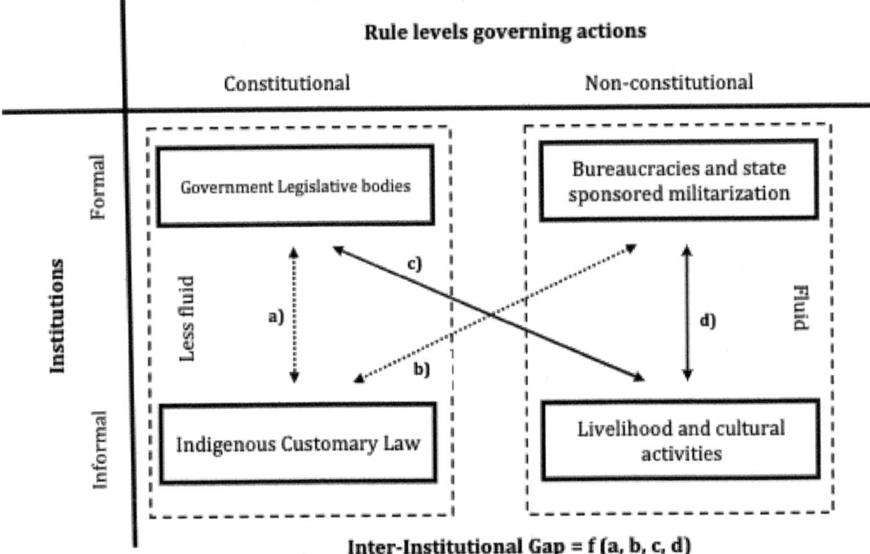

Figure 2: IIG in the Philippines Mining Sector. Adapted from Rahman et al. (2017): A framework for analyzing institutional gaps in natural resource governance.

A. *Legal pluralism: Gap Between Government's Legislative Bodies and Indigenous Customary Law*

Legal pluralism in Gap A of the IIG framework emerges when formal constitutional rules do not value informal constitutional rules governing the same resource system. Rahman et al. (2017) state that legal pluralism results from a lack of understanding that law originates not only from formal governments but from customs and culture as well. In the Philippines' extractive industry, a gap exists between government legal systems and Lumad customary laws.

The impact of legal pluralism is observed in mutual interactions between the industry and political leadership. Specifically, political authorities develop rent-seeking rules by creating formal constitutional choice rules that enhance private interests in mining. Centralized government policies have devalued the importance of Indigenous customary laws, ultimately infringing on the Lumad people's inherent rights to ancestral domains (IIG A). This is done through the process of accommodating informal institutions, defined by Helmke and Levitsky (2004) as "informal institutions create incentives to behave in ways that alter the substantive effects of formal rules, but without directly violating them; they contradict the spirit, but not the letter, of the formal rules." In this case, informal institutions have become very limited in the processes of accommodation. This is seen in formal constitutional choice rules implemented by government legal systems such as the Indigenous Peoples Rights Act that limit and constrain informal constitutional choice rules of Lumad peoples by creating barriers to autonomy and self-determination over ancestral domains.

The accommodation of informal institutions is upheld through the legal pluralism and uncertainty of Philippine state policies. This can be seen in the Philippines 1997 Indigenous Peoples Rights Act (IPRA) and its conflict with other land-use legislation such as the 1995 Philippine Mining Act and Regalian Act, among others. IPRA is known as one of the world's leading progressive legislatures on Indigenous people's cultural and territorial rights. However, the effectiveness of the law's implementation remains problematic due to weak state institutions and political dynasties continuing to wield extensive power in how legislation is carried out at the national and local levels (Doyle 2020).

Moreover, loopholes exist in the constitutional IPRA that contradict other constitutional precepts (IWGIA 2012). For example, the IPRA's recognition of Indigenous ownership of ancestral lands contradicts the standing Regalian Doctrine, which claims all lands are owned by the state as "public domain" (Arellano Law Foundation 2000; Simbulan 2016). Due to the legacy of colonial legal systems, the wholesale incorporation of Indigenous land is common in most post-colonial countries (Rahman et al. 2014).

To challenge the colonial land-rights system under the Regalian Doctrine, the Cariño Doctrine was simultaneously implemented to legally define ancestral domains as private property to Indigenous peoples (LRC-KSK 2001). However, Lumad populations continue to face challenges regarding land ownership and customary rights even under the Cariño Doctrine and IPRA (Espiritu 2017; IWGIA 2012). IPRA enforces limited ownership rights to Indigenous people (Doyle 2020). The Act does not consider Indigenous peoples to be owners of natural resources but rather managers, stewards, and beneficiaries thereof (Arellano Law Foundation 2000). Moreover, legal ownership is only recognized through the government's provision of a "Certificate of Ancestral Domain Title," referred to as CADT (LRC-KSK 2001). In this case, the government's formal codification of Indigenous ownership has infringed on Indigenous rights to self-determination and ancestral domains.

With legal uncertainty across formal constitutional rules, powerful states, with the support of private actors, have used the Regalian Doctrine to reinforce land rights within the 1995 Philippine Mining Act (Arellano Law Foundation 2000). Legal pluralism and uncertainty explain the formal institutional mechanisms that have enabled over 60 percent of mining corporations to operate in ancestral domains (Simbulan 2016). The encroachment of private enterprises onto ancestral territories emerges from heterogeneous and contradictory understandings of land rights and the neglect of Indigenous customary law (Hagen and Minter 2019; IWGIA 2012;

Paredes 2016). The legal uncertainties in formal constitutional rules allow large-scale mining policies to assert state and private enterprise ownership of ancestral lands (Arellano Law Foundation 2000). Legal pluralism has resulted in formal constitutional rules, such as the IPRA, failing to protect and support Lumad self-determination.

B. Cultural Mismatch: Gap Between Indigenous Customary Law, State Militarization and Bureaucracies

Gap B of the IIG framework describes cultural mismatch as a gap that emerges when formal non-constitutional actors (e.g., bureaucrats) develop operational rules without recognizing or understanding the cultural norms of informal constitutional rules (Rahman et al. 2019). In this case, an inter-institutional gap is identified between bureaucrats of different state agencies, the Armed Forces of the Philippines, and Indigenous customary laws.

A cultural mismatch emerges from the dominance of Eurocentric knowledge and power systems in the Philippines and the resulting subjugation of Indigenous worldviews. Paredes (2016) claims that the clash in worldviews between Indigenous and non-Indigenous peoples is reinforced by other religious beliefs where Christian Filipinos consider Indigenous peoples to be primitive and backward. Another cultural mismatch identified in the case is related to the passing of laws that have been copied by procedures used in developed countries. Unsurprisingly, outside the Philippines, the passing of such laws, such as the Indigenous Peoples Rights Act, has been identified as a "success" despite weak enforcement leading to a failure in effective implementation (Andrews et al. 2017).

Moreover, the coercion used by the Armed Forces of the Philippines to protect private industries and the associated lack of respect for Lumad customary laws has further widened this gap. Lumad customary law strongly adheres to the belief that land is life and that, as such, it is not theirs to give away or to sell but to protect and nurture (Alejo 2018; Hagen and Minter 2019; Paredes 2015).

Contrary to Indigenous worldviews, bureaucrats and the Armed Forces of the Philippines view rights to land as defined strictly by national laws that were put in place by landed elites and political dynasties, with communities possessing little power or control in governance matters (Holden and Jacobson 2013; Kreuzer 2009). States view property as a factor of production, either disposable or an investment to generate profit. The state's relationship with the land has led to land-use policies such as the Philippine Mining Act overriding legislation that aims to safeguard Indigenous lands and rights. The Armed Forces of the Philippines and the Philippine National Police, as a national institution, view land through the lenses of the government's established norms in protecting M4D policies (Crost and Felter 2016; Holden et al. 2011; Kreuzer 2009; McCoy 1994).

The reductionist underpinnings of government policies have led to widespread displacement, alongside the oppression of Lumad communities and the marginalization of Lumad worldviews and customary laws. Those implementing mining for development policies do not consider that the forcible removal of Lumad peoples from their ancestral domains has an incalculable impact on their ways of living—one that risks erasing their culture, livelihoods, and existence. According to B'laan belief, which is one of the 18 tribes of the Lumad peoples, deceased community members must be protected within resting places defined by the B'laan as sacred cultural sites (Hamm et al. 2013). Research undertaken by Hamm et al. (2013) indicated that the Tampakan Copper-Gold Mine proposed by Sagittarius Mines Inc. (SMI) violated the integrity of sacred places by bulldozing burial grounds, which led to violent conflicts in the B'laan area. Further, state-sponsored militarization of the communities within the Tampakan area has caused high psychological pressures and violations of human rights related to the life, liberty, and security of the affected community members (Hamm et al. 2013).

IIG Gap B in the Philippines' mineral governance also exists because of competing informal institutions, a common characteristic of clientelism, patrimonialism, and corrupt governance (McCoy 1994). Helmke and Levitsky (2004) define competing informal institutions as "institutions that co-exist with ineffective formal institutions. In such cases, formal rules and procedures are not systematically enforced, which enables actors to ignore or violate them." Competing informal institutions have permitted state actors with strong ties to the Filipino mining industry to violate state laws with impunity, allowing bureaucrats to serve personal interests rather than the public (Acheson 2006; Helmke and Levitsky 2004). As a result, bureaucrats have little incentive and motivation to understand and communicate with informal, non-constitutional rules.

C. Structural Hole: Gap Between State-Sponsored Militarization, Bureaucracies and Indigenous Livelihood Activities

Inter-institutional Gap C emerges when formal constitutional rules deem the actions undertaken by informal non-constitutional rules as illegal and punishable by law (Rahman et al. 2019). Structural holes emanate from this gap when formal and informal actors with no direct network are influenced by a third actor (Rahman et al. 2017). In Gap C, the formal and informal institutions include Philippine bureaucracies and Indigenous livelihood activities, respectively. The third actor in this structural hole is local government officials who are often connected to powerful oligarchies and might possess private interests in mining corporations.

Structural holes have led constitutional formal actors to characterize anti-mining Lumad communities as terrorists and rebels by anti-communist M4D policymakers (Holden 2014; Kreuzer 2009; McCoy 1994). In IIG C, nationalist M4D discourse demonizes Indigenous activism, resulting in red tagging and state violence against land defenders (Tugade 2020). This gap has further enabled elite government officials to undertake rent-seeking activities.

Political and influential leaders who are well-networked with the extractive industry use their power and capacity within government bureaucracy in their favor. Local government officials are commonly connected to the elite family oligarchies who dominate the nation's political arena (McCoy 1994). These third-party actors can be found at the regional, provincial, or city level (Lindon et al. 2014). To protect private interests, bureaucrats can influence state-sponsored militaries to act violently toward Indigenous activists as a means of political suppression against dissent (Joaquin and Biana 2020).

The recent passing of the Anti-Terrorism Act on June 3, 2020, is exemplary of potential structural holes in governance. This Act created an Anti-Terrorism Council, which is made up of government officials who possess the power to define who terrorists are. It allows the surveillance of suspected individuals for extended periods and allows warrantless arrests and detentions for up to 30 days (Joaquin and Biana 2020). Under government legislation, pro-mining politicians appointed to the Anti-Terrorism Council are in a position of defining and imprisoning anti-mining activists as terrorists committing treason.

Yet, Lumad communities have shown increasing resistance to large-scale mining policies through social movements (Espitiru 2017; Karapatan 2014). For example, the Lumad peoples' organization, KALUMARAN (Katawhang Lumad sa Mindanao), spearheads resistance to mining projects that threaten ancestral lands (Simbulan 2016). Nonetheless, calls for justice and self-determination among Lumad communities have been faced with threats and human rights abuses from government militaries and paramilitaries.

With growing opposition to mining, structural holes, and corruption have become critical to the Philippines' advancement of extractive operations. Already by 2010, state-sponsored military groups were the cause of over 60 deaths of Indigenous leaders in Mindanao (Espiritu 2017). For example, on April 7, 2019, Datu Kaylo Bontolan, a Manobo tribe leader, a protector of ancestral lands, and a strong advocate for Indigenous self-determination, was killed during a military bombardment in Kitaotao, northern Mindanao and later framed as a political criminal by state and armed forces (Global Witness 2020).

D. Institutional Void: Gap Between Indigenous Livelihood Activities and Government Bureaucracy

Inter-institutional Gap D is a result of an institutional void. Institutional voids are known as conflicts derived from a lack of harmonization in agreed-upon rules in resource governance (Rahman et al. 2019). In our case study, an institutional void exists at the nexus of Lumad livelihood activities and national government legislation. In IIG D, institutional voids appear when government bodies illegitimatize Indigenous livelihood activities due to clashing worldviews and differing relationships with the land (Alejo 2018; Espiritu 2017). This governance gap results in government legislation infringing on the Lumad people's right to self-determination.

The informal, non-constitutional rules of Lumad peoples are important components of livelihood activities embedded in a cultural, spiritual, and economic dependence on nature (Simbulan 2016). For example, Paredes (1997) states that the Lumad tribe, Higaunons, "have a moral economy, wherein they respect each other and protect the forest." Indigenous livelihood activities seek to preserve the legacies of ancestral domains maintained across generations (PNFSP 2014). In Lumad's worldview, land is a communal property and the source of livelihood for past, present, and future generations. Today, customary law is taught to youth through Lumad schools, which aim to strengthen inter-generational traditions and culture (Espiritu 2017). Lumad schools, known as the Alternative Learning Center for Agricultural and Livelihood Development (ALCADEV), are volunteer-run and provide secondary education to Lumad youth related to numeracy, literacy, farming, cultural and heritage activities, and customary laws (Santos 2017).

At the opposite end of the spectrum, mainstream government actors view the land from an extractive capitalist perspective, with reductionist definitions essentially organized around individual private land ownership (PNFSP 2014). Worldviews of many government bureaucrats, particularly those working for agencies close to agro-business, mining, and industrial activities and often trained within Western paradigms, are embedded in protecting mining projects and private properties. This commitment to private extraction includes diminishing dissent and, in worse situations, directly eliminating those holding views in opposition to the state (Crost and Felter 2020). Alejo (2018) states that the "simplification and standardization" of Philippine state laws have challenged the Lumad people's rights to self-determination. The author highlights the reductionist underpinnings of government policies by suggesting that laws lack inclusivity with regard to Lumad worldviews and customary practices. This has resulted in infringements on community agencies and a reduced capacity for Lumad peoples to engage in local decision-making. Vivoda (2008) corroborates these findings by stating that regulatory agencies governing foreign mining investment in the Philippines "fail to serve the public interest."

Conclusion

With the Duterte regime (elected in May 2016), the Philippine government once again champions mining for development and large-scale mining policy instruments for its ability to catalyze growth in the archipelago. However, oligarchies of political elites and transnational mining corporations sometimes enable fuel corruption across the nation's extractive sector, limiting potential economic outcomes and exacerbating negative externalities. In our analysis, we mobilized the IIG framework to explore different co-existing gaps in natural resource governance. As shown, many have resulted in Philippine state-sponsored repression and militarization in Lumads' ancestral domains. In this chapter, we focused particularly on institutions governing mineral resource extraction, namely government bureaucracies, state militarization, Indigenous customary law, and Indigenous livelihood activities. The identified governance gaps have been classified into four types: legal pluralism, cultural mismatch, structural holes, and institutional voids:

- Legal pluralism results in government legislation systematically failing to protect and value Lumad peoples' inherent rights to their ancestral domain;
- Cultural mismatch is a consequence of Eurocentric dominance of knowledge systems and subjugation of Indigenous worldviews in mineral governance. Extractivist and productivist worldviews have enabled and sustained large-scale mining policies with political elites using state violence to protect private industries while abusing Lumad customary laws and human rights;
- Structural holes have led to the interpretation of Indigenous activism as terrorism and justified political suppression of dissent;

- Lastly, institutional voids emerge from government legislation that does not recognize Lumad self-governing activities and considers them illegitimate and threatening. This chapter provides a foundation for future research and policy perspectives to create avenues for peacebuilding between governments and anti-mining Lumad communities. Policy changes for mitigating conflicts will require restructuring the legal framework within the Philippines in order to allow legal recognition and implementation of Indigenous governance practices in the management of protected areas and ancestral domain land-use plans. In support of this process, action-oriented research could examine legal mechanisms to uphold inter-institutional accountability and eliminate corporate impunity for human rights violations in the Philippines' extractive industry.

Lessons, Limitations, and Opportunities

The IIG framework was originally developed to address site-specific governance issues; therefore, our application of the conceptual tool on a national-level case study posed methodological challenges. Important data deficits, including the role of peripheral agents influencing inter-institutional arrangements, limited the depth and scope of the study.

Importantly, the case study was not informed by primary data collection. Our reliance on secondary data may have introduced inaccuracies with regard to our representations of Lumad institutions and their associated rules, worldviews, and social habits. This limitation may have resulted in an overly generalizing account of anti-mining Lumad communities. Through the collection of disaggregated data between tribes, future studies might allow for more profound and nuanced understandings of the discrete realities Indigenous communities in Mindanao face with large-scale mining operations.

This study may also be limited by our choice to ignore the diverse roles peripheral agents may play in natural resource governance. NGOs, business groups, and political and environmental activists exert considerable positive and negative pressure upon these systems, which is difficult to capture by our analytical approach. Strong monitoring of the conflicting interests of different rule-developing agents is imperative for sustainable natural resource management.

Future action-oriented research could examine avenues that eliminate corporate impunity for human rights violations in the Philippines. Policy-driven studies could aim to identify and strengthen legal mechanisms and rights-based approaches to support local capacity-building initiatives to resolve conflict situations peacefully. Capacity-building initiatives could help empower local and Indigenous communities with knowledge related to the 1997 Indigenous Peoples Rights Act, the 1995 Philippine Mining Act, mining concessions, royalty and tax payments, etc. This initiative would mobilize and share knowledge related to mining and legal technicalities with local communities, strengthening local participation in resource governance while facilitating dialogue between communities, governments, and companies to promote transparency and accountability within the sector. Some of these ideas are currently being undertaken by NGOs such as Bantay Kita, a national coalition advocating for community empowerment and meaningful local participation in resource governance within the Philippines (Bantay Kita 2021).

Acknowledgments

We want to acknowledge with extreme gratitude the late Dr. Nonita Yap, Dr. John Devlin, and the Global Minerals Local Communities in Canada and the Philippines research team for their ongoing support and review of this chapter. This project was supported financially by the Social Sciences and Humanities Research Council, the School of Environmental Design and Rural Development, and the Rural Planning and Development Departments at the University of Guelph.

References

Acheson, J.M. 2006. Institutional failure in resource management. Annual Review of Anthropology 35(1): 117–134. doi:10.1146/annurev.anthro.35.081705.123238.

Alejo, A.E. 2018. Strategic identity: Bridging self-determination and solidarity among the indigenous peoples of Mindanao, the Philippines. Thesis Eleven 145(1): 38–57. doi:10.1177/0725513618763839.

Andrews, M., M. Woolcock and L. Pritchett. 2017. Building State Capability: Evidence, Analysis, Action. Oxford University Press.

Arellano Law Foundation. 2000, December. Ikalahan Indigenous People and Haribon Foundation for the Conservation of Natural Resources, Inc. Retrieved from https://lawphil.net/judjuris/juri2000/dec2000/gr_135385_2000.html.

Aytin, A. 2015. A Social Movements' perspective on human rights impact of mining liberalization in the Philippines. NEW SOLUTIONS: A Journal of Environmental and Occupational Health Policy, 25(4): 535–558. doi: 10.1177/1048291115608354.

Bantay Kita. 2021. History. Retrieved from http://www.bantaykita.ph/bkhistory.html.

Broad, R. and J. Fischer-Mackey. 2016. From extractivism towards buen vivir: Mining policy as an indicator of a new development paradigm prioritising the environment. Third World Quarterly, 38(6): 1327–1349. doi:10.1080/01436597.2016.1262741.

Camba, A.A. 2015. From colonialism to neoliberalism: Critical reflections on Philippine mining in the "long twentieth century". The Extractive Industries and Society 2(2): 287–301. doi:10.1016/j.exis.2015.02.006.

Camba, A.A. 2016. Philippine mining capitalism: The changing terrains of struggle in the neoliberal mining regime. ASEAS – Austrian Journal of South-East Asian Studies 9(1): 69–86.

Caouette, D. 2012. Oligarchy and caciquismo in the Philippines. pp. 169–180. *In:* Neopatrimonialism in Africa and Beyond. Routledge. https://doi.org/10.4324/9780203145623–19.

Ciencia, A.N. 2011. The Philippine Supreme Courts Ruling on the Mining Act: A Political Science Perspective. Philippine Political Science Journal 32(55): 1–36. doi:10.1080/01154451.2011.9723530.

Crost, B., J.H. Felter and P.B. Johnston. 2016. Conditional cash transfers, civil conflict and insurgent influence: Experimental evidence from the Philippines. Journal of Development Economics 118: 171–182. https://doi.org/10.1016/j.jdeveco.2015.08.005.

Crost, B. and J.H. Felter. 2020. Extractive resource policy and civil conflict: Evidence from mining reform in the Philippines. Journal of Development Economics 144: 102443 doi:10.1016/j.jdeveco.2020.102443.

Doyle, C. 2020. The Philippines Indigenous Peoples Rights Act and ILO Convention 169 on tribal and indigenous peoples: exploring synergies for rights realisation. The International Journal of Human Rights: ILO Convention 169: Critical Perspectives 24(2-3): 170–190. https://doi.org/10.1080/13642987.2019.1679120.

Escobar, A. 2012. Encountering development: The making and unmaking of the Third World. Princeton: Princeton University Press.

Espiritu, B.F. 2017. The Lumad Struggle for Social and Environmental Justice: Alternative Media in a Socio-Environmental Movement in the Philippines. Journal of Alternative & Community Media 2(1): 45–59. doi:10.1386/joacm_00031.

Global Witness. 2020. Defending Tomorrow: The climate crisis and threats against land and environmental defenders. Retrieved from https://www.globalwitness.org/en/campaigns/environmental-activists/defending tomorrow/.

Graulau, J. 2008. 'Is mining good for development?' Progress in Development Studies 8(2): 129–162. doi:10.1177/146499340700800201.

Hagen, R.V. and T. Minter. 2019. Displacement in the name of development. How indigenous rights legislation fails to Protect Philippine Hunter-Gatherers. Society & Natural Resources 33(1): 65–82. doi:10.1080/08941920.2019.1677970.

Hamm, B., A. Schax and C. Scheper. 2013. Human Rights Impact Assessment of the Tampakan Copper-Gold Project. Institute for Development and Peace, University of Duisburg Essen.

Helmke, G. and S. Levitsky. 2004. Informal institutions and comparative politics: a research agenda. Perspectives on Politics 2(04): 725–740. doi:10.1017/s1537592704040472.

Holden, W. and Jacobson, Dan. 2007. Mining amid armed conflict: Nonferrous metals mining in the Philippines. The Canadian Geographer/Le Géographe canadien. 51: 475–500. 10.1111/j.1541-0064.2007.00193.x.

Holden, W., K. Nadeau and R.D. Jacobson. 2011. Exemplifying accumulation by dispossession: mining and indigenous peoples in the philippines. Geografiska Annaler: Series B, Human Geography 93(2): 141–161. doi: 10.1111/j.14680467.2011.00366.x.

Holden, W.N. and R.D. Jacobson. 2013. Mining and natural hazard vulnerability in the philippines: digging to development or digging to disaster? London: Anthem Press.

Holden, W.N. 2014. The New Peoples Army and Neoliberal Mining in the Philippines: A Struggle against Primitive Accumulation. Capitalism Nature Socialism 25(3): 61–83. doi: 10.1080/10455752.2014.922109.

International Work Group for Indigenous Affairs (IWGIA). 2012. Philippines: Obstacles concerning access to justice and protection for indigenous people. Retrieved from https://www.iwgia.org/en/philippines/1678-philippines-obstaclesconcerning-access-to-justice.html.

Iglesias, S. 2018. Central-Local Dynamics and Political Violence in the Philippines, 2001 to2016 [ProQuest Dissertations Publishing http://search.proquest.com/docview/2190677716/.

Joaquin, J.J. and H.T. Biana. 2020. Philippine crimes of dissent: Free speech in the time of COVID-19. Crime, Media, Culture: An International Journal, 174165902094618. doi:10.1177/1741659020946181.

Karapatan. 2014. Karapatan Year-end Report on the Human Rights. Retrieved from https://www.karapatan.org/2014 Human Rights Report.

Khee-Jin Tan, A. 2006. All that glitters: foreign investment in mining trumps the environment in the Philippines. Pace Environmental Law Review 23(1): 182–208. doi: 10.1017/cbo9780511542169.028.

Kreuzer, P. 2009. Private political violence and boss-rule in the Philippines. Behemoth 2(1). doi:10.1524/behe.2009.0005.

Legal Rights and Natural Resources Center (LRC-KSK). 2001. A Divided Court, A Conquered People? Case Materials from the Constitutional Challenge to the Indigenous Peoples' Rights Act of 1997. Quezon City: Legal Rights and Natural Resources Centre Inc. Manila: Kasama sa Kalikasan LRC-KSK, Friends of the Earth-Philippines, 15.

Lindon, J., R.U. Mendoza and T.A. Canare. 2014. Governance and Market Failures in Mining: Lessons from the Marcopper Mine Disaster in Marinduque, Philippines. SSRN Electronic Journal. doi:10.2139/ssrn.2345228.

Magno, C. 2015, August 31. Fair Compensation and other Prerequisites to Mining for Development. Retrieved from http://www.unrisd.org/road-to-addis-magno.

McCoy, A. 1994. An anarchy of families state and family in the Philippines. Philippine Political Science Journal 18(1-2): 149–153. doi:10.1163/2165025x-0180102007.

Meger, S. and J. Sachseder. 2020. Militarized peace: understanding post-conflict violence in the wake of the peace deal in Colombia, Globalizations 17: 6, 953973, DOI: 10.1080/14747731.2020.1712765.

Ocampo, K., L.B. Salaverria and N. Corrales. 2021, April 16. Duterte lifts 9-year ban on new mining deals. Retrieved from https://newsinfo.inquirer.net/1419780/duterte-lifts-9-yearban-on-new-mining-deals.

Paredes, O. 1997. People of the Hinterland: Higaûnon Life in Northern Mindanao, Philippines. Unpublished M.A. Thesis, Arizona State University.

Paredes, O. 2015. Indigenous vs. native: Negotiating the place of Lumads in the Bangsamoro homeland. Asian Ethnicity 16(2): 166–185. doi:10.1080/14631369.2015.100369.

Paredes, O. 2016. Rivers of memory and oceans of difference in the Lumad World of Mindanao. TRaNS: Trans -Regional and -National Studies of Southeast Asia 4(2): 329349. doi:10.1017/trn.2015.28.

Paredes, O. 2019. Preserving 'tradition': The business of indigeneity in the modern Philippine context. Journal of Southeast Asian Studies 50(1): 86+ https://link.gale.com/apps/doc/A583488885/AONE?u=guel77241&sid=bookmarkAONE&xid=3002b1a6.

Pedro, A., E.T. Ayuk, C. Bodouroglou, B. Milligan, P. Ekins and B. Oberle. 2017. Towards a sustainable development licence to operate for the extractive sector. Mineral Economics 30(2): 153–165. doi:10.1007/s13563-017-0108-9.

Philippine Network of Food Security Programes (PNFSP). 2014, June 05. The Philippine Indigenous People and their Customary Laws. Retrieved from https://www.slideshare.net/PNFSP/pnfsp-ip-customarylaws.

Rahman, H.M.T., S.K. Sarker, G.M. Hickey, M.M. Haque and N. Das. 2014. Informal institutional responses to government interventions: lessons from Madhupur National Park, Bangladesh. Environmental Management 54(5): 1175–1189. doi:10.1007/s00267014-0325-8.

Rahman, H.M.T., A.S. Saint Ville, A.M. Song, J.Y.T. Po, E. Berthet, J.R. Brammer et al. 2017. A framework for analyzing institutional gaps in natural resource governance. International Journal of the Commons 11(2): 823–853. DOI: http://doi.org/10.18352/ijc.758.

Rahman, H.M.T., J.Y. Po, A.S. Ville, N.D. Brunet, S.M. Clare, S. Darling et al. 2019. Legitimacy of different knowledge types in natural resource governance and their functions in inter-institutional gaps. Society & Natural Resources 32(12): 1344–1363. doi:10.1080/08941920.2019.1658140.

Rodil, R.B. 2020. Kasaysayan ng mga pamayanan ng Mindanao at arkipelago ng Sulu, 1596–1898. Maynila: Pambansang Komisyon para sa Kultura at mga Sining.

Rovillos, R.D. and V. Tauli-Corpuz. 2012. Development, Power, and Identity Politics in the Philippines. The Politics of Resource Extraction. doi:10.1057/9780230368798.0012.

Santos, J. 2017. For Lumad schools, even holding class is a struggle. Indigenous Peoples Major Group For Sustainable Development. Retrieved from https://indigenouspeoplessdg.org/index.php/english/ttt/771-for-lumad-schools-even-holding-class-is-a-struggle.

Simbulan, R.G. 2016. Indigenous Communities' Resistance to Corporate Mining in the Philippines. Peace Review 28(1): 29–37. doi:10.1080/10402659.2016.1130373.

Tugade, R.R. 2020. The Modern Scarlet Letter: Red-Tagging of Civilians as Violation of the Principle of Distinction in International Humanitarian Law. 10.13140/RG.2.2.14035.60962.

Vivoda, V. 2008. Assessment of the Governance Performance of the Regulatory Regime Governing Foreign Mining Investment in the Philippines. Minerals & Energy – Raw Materials Report 23(3): 127–143. doi:10.1080/14041040902805165.

Wilson, S. 2019. Militarised security: Understanding the relationship between the armed forces of the Philippines and human rights. Human Rights Defender 28(2): 34–35.

Zech, S.T. and J. Eastin. 2021. The Household Economics of Counterinsurgency. Defence and Peace Economics 32(2): 220–239. doi:10.1080/10242694.2019.168859.

4

Tiger Conservation Governance in the Bangladesh Sundarbans
Identifying Inter-Institutional Gaps

Swapan Kumar Sarker,[1,*] *Md. Modinul Ahsan,*[2]
A. Z. M. Manzoor Rashid,[1] *Abu Naser Mohsin Hossain,*[3]
Md. Bashirul Al Mamun[4] *and H. M. Tuihedur Rahman*[5]

Introduction

The global tiger habitat range has declined by 93 percent, with only 3,500–4,000 tigers surviving in 13 Tiger Range Countries (TRCs) (Aziz 2017; Walston et al. 2010; Khan et al. 2018). South Asia supports about 60 percent of the global tiger population despite experiencing the highest level of tiger population loss (98 percent) in the last 200 years (Aziz 2017). Habitat degradation, prey scarcity, poaching, and illegal trade are some of the main anthropogenic challenges to tiger conservation (Mondol et al. 2009; Inskip et al. 2013; Saif et al. 2018). The TRCs have taken drastic efforts to design and implement a variety of institutional measures and spend around USD 50 million to support the *in situ* conservation of tigers in the remaining 7 percent of

[1] Department of Forestry &, Environmental Science, Shahjalal University of Science & Technology, Kumargaon, Sylhet - 3114, Bangladesh.
  Email: pollen_forest@yahoo.com
[2] USAID Ecosystems/Protibesh Activity, Chemonics International Inc. Gulshan 2, Dhaka - 1212, Bangladesh.
  Email: modinul2704@gmail.com
[3] Sundarban West Forest Division, Bangladesh Forest Department, K.D. Ghos Road,Khulna - 9100, Bangladesh.
  Email: abunasermh@gmail.com
[4] Dhaka Forest Division, Bangladesh Forest Department, Bangladesh.
  Email: mamun98sust@yahoo.com
[5] Department of Agricultural and Resource Economics, University of Saskatchewan, 51 Campus Dr, Saskatoon, Saskatchewan, Canada S7N 5A8.
  Email: hm.rahman@mail.mcgill.ca
* Corresponding author: swapan-fes@sust.edu

tiger habitat (1,185,000 km^2) (Wilting et al. 2015). Common measures taken by TRC governments include enacting new wildlife laws, protecting tiger habitats through establishing new protected areas and enforcing laws to stop poaching (Rastogi et al. 2014; Johnson et al. 2018; Richardson et al. 2020). Despite these reforms, existing conservation institutions have been unable to address some of the major challenges for tiger conservation in many critical ecosystems, including the Sundarbans Mangrove Forest (SMF) of Bangladesh—the only mangrove tiger habitat in the world (Aziz 2017). One common challenge with formal institutional measures is that they need to account for the local social and cultural contexts that determine the nature of human-wildlife interactions.

Influenced by a wide range of cultural, economic, social, and political forces, government agencies and resource-using communities often develop various forms of mutually agreed-upon rule systems known as institutions (North 1991; Broad and Damania 2010). Codified rules that act as a *de jure* regulatory mechanism are known as formal institutions and are commonly developed and enforced by a hierarchically organized rulemaking and enforcement body. For instance, the Bangladesh Forest Department follows the Forest Act of 1927 and the Wildlife (Conservation and Security) Act of 2012 to guide the conservation of tigers and other natural resources in the Sundarbans. On the other hand, informal institutions that commonly function as a *de facto* rulemaking and enforcement system comprise habitual, verbalized, or customary norms, values, and conducts (North 1991). For instance, tiger-related social norms, values, and practices, in addition to local political factors, form the informal institutions in the SMF. Both formal and informal institutions are arranged into three layers of "rules-in-use" (also called rule levels): operational choice, collective choice, and constitutional choice rule levels (see Chapter 2 for details). Rahman et al. (2017) suggest that a lack of coordination between the rule levels of formal and informal institutions is likely to occur if they operate in a single resource system because of the diversity of institutional operations and objectives. This lack of coordination may result in inter-institutional gaps (IIGs), and if they remain unnoticed and unaddressed, the IIGs may result in conflicting interactions between governments and resource-using communities (Rasul 2007; Rahman et al. 2014) Thus, the IIG framework recategorizes the level of rules into constitutional (i.e., constitutional choice) and non-constitutional (i.e., operational and collective choice) level rules to analyze the interactions between different levels of formal and informal institutions (Rahman et al. 2017).

Bengal tigers have a strong socio-cultural influence on the social-ecological system in the SMF (Khan et al. 2018). As top predators, they play a central role in maintaining SMF's ecosystem services and functions (BFD 2016a). Due to its cultural and ecological significance, the animal has led biodiversity conservation campaigns in Bangladesh. Although multiple local, national, and international actors are involved in tiger conservation in the SMF (Barlow et al. 2008; Aziz 2017), a clear understanding of the interplay between the formal and informal institutions in conserving tigers is lacking. The declining number of tigers, coupled with increasing human-tiger conflicts and livelihood insecurity among mangrove resource-dependent communities, has created a need to better understand how the country's forest department, policy-making bodies, local communities, and non-government organizations can jointly work to intensify conservation efforts. Using multiple methods appropriate for applying the Inter-Institutional Gaps (IIGs) framework (Rahman et al. 2017), this chapter aims to identify gaps between the formal and informal institutions involved in tiger conservation in the SMF. Our specific questions include:

(I) How do different constitutional and non-constitutional rule levels (under formal and informal institutions) interact with each other?

(II) Is there any interactional gap (i.e., IIGs) occurring between the rule levels operating under the formal and informal institutions?

Overall, this chapter intends to help better understand how the interactions between formal and informal institutions should be studied in the context of wildlife conservation.

In what follows, we discuss human-forest-tiger interactions in the SMF, followed by a general description of the study system. We, therefore, explain data collection and analysis methods following the structure of the IIG framework before presenting results. Finally, we discuss and conclude the implications of our results for sustainable tiger conservation in the Sundarbans.

A Brief Overview of the Sundarbans Mangrove Forest (SMF)

The Sundarbans is the largest continuous mangrove forest in the world, covering 10,017 km² in Bangladesh and India. The Bangladesh part of the Sundarbans (i.e., SMF, Figure 1) possesses a 6,017 km² area with 69 percent forest land and the

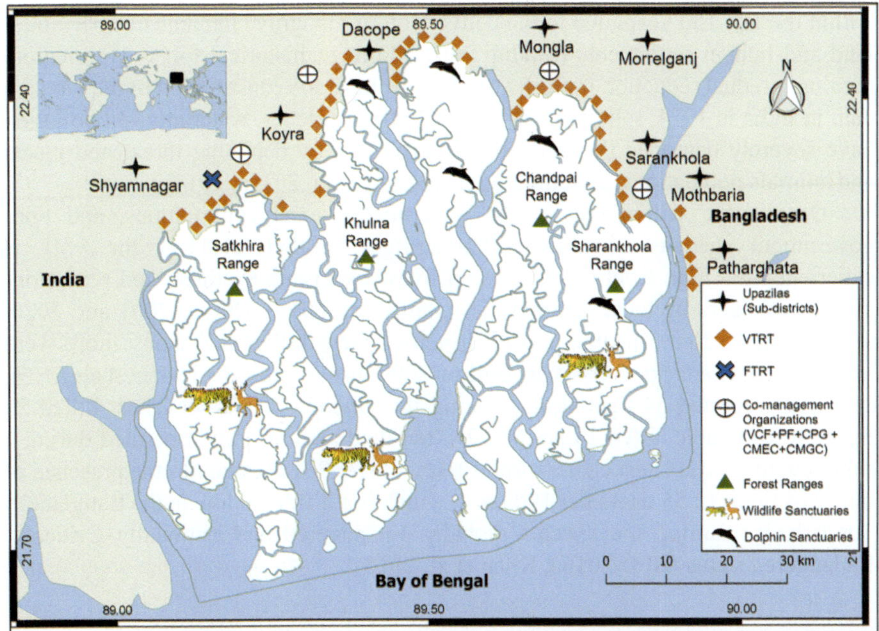

Figure 1: The Sundarbans Mangrove Forest (SMF) with spatial distributions of the actors involved in tiger conservation under formal and informal institutions.

rest encompassing rivers, small streams, and canals. The SMF is a Level I Tiger Conservation Unit with global importance and supports the Earth's remaining Bengal Tiger population (Khan 2004; Khan et al. 2018). The government of Bangladesh has established three wildlife sanctuaries (3,180 km^2) comprising 52 percent of the area of the SMF (Aziz and Paul 2015; Sarker et al. 2019a). SMF is predominantly managed by the Bangladesh Forest Department (BFD) under three Divisional Forest Offices (DFOs), including Sundarban East, Sundarban West, and the Wildlife Management and Nature Conservation Divisions of Khulna (WMNCDK). The SMF is divided into four Forest Ranges (i.e., Chandpai, Sharankhola, Satkhira, and Khulna), 16 Forest Stations, 64 Forest Camp/Petrol offices, and 55 management compartments. Local BFD officials (e.g., forest rangers, foresters, forest guards, and boatmen) operate these offices to conduct regular forest management and conservation activities (BFD 2019). Also, different national and international NGOs, co-management organizations, and donor agencies play important roles in tiger conservation (Barlow et al. 2008; Aziz 2017).

While there is no permanent human settlement inside the forest, the SMF is surrounded by 76 villages (except the Bay of Bengal coast in the south) and more than 3.5 million people comprising fishers, woodcutters, honey, and other non-wood forest product harvesters directly or indirectly depend on the SMF's resources (Sarker et al. 2016). Most of these resource users are socio-economically vulnerable due to their low income and unstable livelihood opportunities (Islam et al. 2014). The SMF also harbors breeding and nursing grounds for numerous marine organisms, including globally endangered plant and animal species (Sandilyan and Kathiresan 2012; Siddiqi 2001), and gives protection from natural disasters like cyclones and oceanic surges (Danielsen et al. 2005). However, the SMF has lost half of its original size within the last 150 years due to the conversion of mangrove habitats to agricultural land and human settlements (Siddiqi 2001). Further, historical forest exploitation, hunting, gradual reduction in freshwater flows due to the construction of the Farakka dam in India in 1974, salinity intrusion, oil spills, cyclones, water and soil pollution have severely degraded the Sundarbans ecosystem by depleting threatened plants and animals populations, including tigers (Sarker et al. 2019a, 2019b).

Despite the willingness of the government, local communities, and non-government agencies to mitigate human-tiger conflicts (HTCs), the SMF is experiencing a high level of human and livestock killing by tigers and retaliatory killings of tigers by local communities (Saif et al. 2018). During 2001 and 2020, approximately 20–50 people and 80 livestock were killed, and many more were injured by tiger attacks every year (Barlow 2009; Inskip et al. 2013; Aziz et al. 2013). The BFD reported 54 tiger deaths during this period. Of this, poachers killed 25 tigers, local people killed 14 stray tigers, ten died naturally, and one died during a super cyclone. The latest camera-trap survey (2016–2018) estimates the presence of only 114 tigers (2.55 tigers per 100 km^2), which is a historical low in the Bangladesh Sundarbans, earning the species globally Endangered and nationally Critically Endangered status (BFD 2016a; Khan et al. 2018).

Methods

An institutional system has three rule levels: operational choice (day-to-day practical choice from alternatives), collective choice (who is allowed to make the choices and what rules and strategies are to be used to make the operational choices), and constitutional choice (who are to be empowered to make collective and operational choices) rules (McGinnis and Ostrom 2014). The IIG framework reclassifies these rules into two levels: constitutional (i.e., constitutional choice rules) and non-constitutional choice rules (collective and operational choice rules) (Rahman et al. 2017). Applying the IIG frameworks requires data on five Analytical Criteria (AC): AC1—formal constitutional; AC2—formal non-constitutional; AC3—informal constitutional; AC4—informal non-constitutional; AC5—boundary organizations. Therefore, we collected data from multiple sources: policy documents review and key-informant interviews (Table 1). For Analytical Criteria 1 (AC1: formal

Table 1. Data Requirements for Applying the IIG Framework

Analytical Criteria (AC)	Key Questions	Data Source
1. Formal constitutional	• What are the government's key tiger conservation commitments? • What are the legal guidelines (e.g., acts, policies, international agreements, etc.) for tiger conservation? • Who are the key government actors involved in tiger conservation?	Policy, legal, and international legal documents.
2. Formal non-constitutional	• Who are the government bureaucrats? • What are their key roles? • How do they engage with local communities? • How do they respond to a tiger-related incident? • What are the key challenges of performing bureaucratic roles and responsibilities?	Interviewing government officials and legal documents.
3. Informal constitutional	• What are the key tiger-related cultural practices? • What are the traditional uses of tiger body parts? • How do the communities relate tigers to their social life? • Is there any tiger-hunting culture present among the community members? • What is the religious and mythological significance of the tiger?	Expert interview, community member interview, and historical document analysis.
4. Informal non-constitutional	• Who are the key informal non-constitutional actors? • Is there any community leadership that responds to a tiger-related incident? • What are the community leadership roles? • How do the communities organize themselves to respond? • Do the community members coordinate with government bureaucrats?	Community member interviews, FGDs, community leader interviews, etc.
5. Boundary organizations	• What are these organizations? • What are their roles? • How do they engage with government and community stakeholders?	Interviews

constitutional), we reviewed forest and tiger conservation-related Bangladesh government's policies, acts, plans, strategies, relevant national and international legal documents. To develop a comprehensive understanding of the existing formal institutional arrangements for tiger conservation in Bangladesh, we analyzed these documents based on key questions identified a priori (Table 1). For AC2 (formal non-constitutional), we interviewed the regional and field-level forest officials responsible for the SMF's overall management [Divisional Forest Officers, i.e., DFOs ($n = 3$), Forest Rangers ($n = 4$), Foresters ($n = 8$) and Forest Guards ($n = 10$)] to understand how tiger conservation rules are being enforced in the SMF. Table 1 presents the questions asked during the interviews. For AC3 (informal constitutional), we interviewed experts with extensive working experience in local culture and tradition ($n = 2$) and local community members ($n = 20$). In addition, we analyzed historical documents based on the questions presented in Table 1 to understand the evolution of local traditional or customary laws related to tiger conservation. For AC4 (informal non-constitutional), we interviewed the same respondents of AC3. We also conducted four focus group discussions (FGDs) in four Forest Ranges. FGD participants included forest resource users, hunters, and tiger victims. The aim of these interviews and FGDs was to understand the informal mechanisms within the communities to respond to tiger-related incidents and to coordinate with forest officials in conserving tigers. For AC5 (boundary organizations), we interviewed the members of three boundary organizations: Village Tiger Response Team (VTRT; $n = 10$), Forest Tiger Response Team (FTRT; $n = 4$), and Collaborative Management Organizations ($n = 5$). During the interviews, we asked them to mention the activities of the organizations in conserving tigers and mediating human-tiger conflicts, as well as their involvement with government and community stakeholders. Data analysis involved the classification of responses on the basis of the questions asked under each analytical criterion (following Elo and Kyngäs 2008).

Results

Overall, the Bangladesh government's HTC management initiatives are driven by two main strategies: first, protecting tigers using legal instruments to stop poaching, enhancing habitat and pray conservation, and building awareness and institutional capacity; and second, protecting people from tiger attacks by enhancing community capacity and providing institutional support, and compensating tiger victims. Communities, on the other hand, have cultural beliefs and behavior toward tigers, which are influenced by a long history of HTCs and the illegal national and international market of tiger body parts. While the government has a mandate for tiger conservation, local communities' culture and behavior are at odds with the government's interventions.

Analysis of Acts, Policies, Plans, Strategies, and International Agreements Related to Tiger Conservation — Formal Constitutional Choice Rules

The government of Bangladesh's acts, rules, and policies have three main purposes: clarifying the legal ownership of the SMF, protecting tigers and their habitat from human interference, and minimizing HTC. In summary:

- The Forest Act, 1927 (Amended in 2000) established complete state control over the rights and ownership of the SMF and its resources. The government also preserves the authority to develop new legal instruments when required (GoB 1927).
- The Forest Policy of Bangladesh, 1994, emphasizes biodiversity conservation and the livelihood security of forest-dependent communities through collaborative forest management.
- The Wildlife (Conservation and Security) Act of 2012 defines the tiger and its main prey—spotted deer—as "protected animals" and increases punishments for tiger killing (GoB 2012).
- The main objective of establishing three SMF wildlife sanctuaries between the 1960s and 1990s under the Bangladesh Wildlife (Preservation) (Amendment) Act of 1974 was mainly to ensure secured territories for tigers (Sarker et al. 2016).
- In 1999, the government declared a 10 km wide strip along the outer side of the SMF (totaling an area of 1,904.5 km²) as an Ecologically Critical Area (ECA) under the Bangladesh Environment Conservation Act, 1995, (GoB 1995) to protect the SMF from externally operated destructive activities. The ECA Management Rules of 2016 restrict tree harvesting, hunting, killing, collecting wildlife, and destroying plants, animals, and aquatic habitats in the ECA.
- The Protected Area Management Rules of 2017 (GoB 2017) delineate the provisions for forming and deploying various community-based platforms to manage and protect mangrove resources by practicing a collaborative governance approach within the periphery of SMF.

Apart from these acts and rules, a number of national and international policies and strategies are in place to conserve tigers (Ahmad et al. 2009; BFD 2010). For example, the Bangladesh Tiger Action Plan (BTAP), 2009–2017 (updated 2018–2027), was developed to reduce HTCs (Khan et al. 2018). The government also has legal provisions to protect and compensate the victims of tiger attacks (i.e., Compensation Rules for the Wildlife Victims, 2010, under the Wildlife (Preservation) (Amendment) Act 1974, and the Compensation Rules for the Wildlife

Victims 2021 under the Wildlife (Conservation and Security) Act 2012). Also, as a part of the Global Tiger Recovery Program (GTRP), Bangladesh prepared the first National Tiger Recovery Program (NTRP) in 2010, revised in 2016, to achieve a demographically stable tiger population in the SMF by 2022 through scientific management, effective conservation programs, institutional capacity building, local community engagement, habitat protection and transboundary collaboration.

Bangladesh Forest Department's Operational Activities for Tiger Conservation: Formal Non-Constitutional Choice Rules

The DFOs of the Sundarban East and Sundarbans West Forest Divisions are responsible for the operational management of the SMF. Most of the local BFD officials have forest management-related training, and their main job responsibilities include regulating resource extraction, issuing forest entrance permits, collecting revenues, and preventing illegal activities. To facilitate wildlife management activities, the BFD established the Wildlife Management and Nature Conservation Division, Khulna (WMNCDK). WMNCDK staff receive training on wildlife monitoring, wildlife crime investigation, and HTC mitigation. However, due to the WMNCDK's insufficient institutional capacity and resources, the DFOs of the Sundarban East and Sundarban West Forest Divisions currently administer the wildlife sanctuaries and tiger conservation activities. During our interview sessions, the BFD officials claimed that experienced and trained WMNCDK staff's frequent job transfer to different locations in the country and lack of legal knowledge deter tiger conservation activities in the Sundarbans (Khan 2004)b.

To minimize HTCs, local BFD officials often get help from the community-based Village Tiger Response Teams (VTRTs), formed by the WildTeam, an NGO working for tiger conservation since 2007. This help is sought when a tiger enters a village. Also, the BFD formed a Forest Tiger Response Team (FTRT) under the Sundarbans Tiger Project 2007. FTRT was a boat-based team responsible for tackling human killing by tigers inside the forest by offering medical assistance, transporting tiger victims, retrieving bodies, and patrolling areas where man-eaters were active. However, the team was limited to the western part of the SMF (particularly in the Satkhira forest range), and the activities of the team stopped in 2018.

Local BFD officials also implement the government's legal provisions to compensate tiger victims. A committee headed by an Upazilla Nirbahi Officer (UNO) (i.e., head of the local civil administration) manages the compensation mechanism. Other members of the committee include a Chairman of a Union Council (i.e., elected representative of the lowest administrative unit of the government), a member of the Union Council, and the Assistant Conservator of Forests or a Range Officer. The committee files a compensation application to a DFO or Warden after a field investigation within 30 days of an incident, and the DFO forwards the application to the Additional Chief Warden (ACW) of the forest with his recommendation within the next 15 days. The ACW verifies the application within seven days and forwards it to the Chief Warden for final budgetary allocation.

Community's Cultural Beliefs, Practices, and Interactions with Tigers: Informal Constitutional Choice Rules

In local folklore, tigers have been described as a fear and foe. Thus, before entering the forest, people make offerings to folk deities and seek blessings from local spiritual/religious leaders, believing that they might help them get rid of tiger attacks. To get protection from tigers, people seek blessings and collect blessed pieces of red cloth and other charms from *Gunins* (local professional spiritual men) as well as from other spiritual leaders (locally known as *Hujur* or *Peer*) before entering the forest. People work in groups in the forest, and some groups (particularly honey collectors) appoint Gunins for a trip to get spiritual protection. A Gunin confirms that group members follow various religious practices to show proper respect to forest spirits. People also put sacred red flags in their working area to avoid tiger attacks and keep close family members with them, believing they might come first if attacked by a tiger.

Tiger hunting was legal in Bangladesh until 1973. Local forest officials and community members (particularly local hunters, i.e., *Sikaris*) used to get rewards for killing tigers during the British and Pakistani periods. Loss of life and livestock due to tiger attacks created a negative community attitude toward tigers. This attitude, along with the formal tiger-killing policies of the previous British and Pakistani regimes, has facilitated the emergence of several social groups (e.g., villagers, poachers, local hunters, i.e., *shikaris*, trappers, and pirates) involved in tiger killing both in villages and in deep forests. Although these social groups have different motives, methods, and networks, killing tigers to supply tiger body parts to emerging local and international markets is a common intention.

People living around SMF use tiger body parts for traditional medicine, culture, and spiritual purposes. They collect tiger parts from killed stray tigers, dead tigers found in deep forests, or from tiger poachers and hunters. Traditionally, people wear lockets or rings containing tiger canines or other teeth or an amulet containing any tiger part, believing that it might help them cure a range of health problems. People rely on *kobiraj* (traditional healer) to determine which tiger parts are useful for what purposes. Sometimes, women eat the soil of tiger pugmarks, believing it to be a contraceptive, while tiger jewelry is used as protection from danger, enemies, or evil spirits. Tiger bones are frequently buried inside houses to get protection from venomous insects. People also believe wearing tiger jewelry increases social esteem.

Community Responses to Tiger Attacks and Compensation Demands: Informal Non-Constitutional Choice Rules

During data collection, local community members informed that villagers act in a group during a stray tiger incident to ensure safety and to avoid punishment for killing the tiger, although some may get injured or killed during the action. Some community members may contact Village Tiger Response Team (VTRT) members, although BFD staff and VTRT members often fail to restrain crowds. Influential community members (including poachers and pirates) who are involved in the local and international trade of tiger body parts also instigate killings. However, if attacked inside the forest, many resource collectors consider it their professional fate. Many tiger victims die immediately after an attack, as nearby BFD posts are often incapable of providing medical assistance. An injured victim may remain untreated until his group members take him to a local hospital, which usually takes a 1–2 day boat journey. Many forest resource users are also killed or injured when they try to save their attacked group member or retrieve the victim's body for burial. In order to notify other forest users about an incident, the attack zone is usually marked with a piece of cloth attached to a tree or pole.

To minimize HTC, local community members have developed several community-based VTRTs with the support of a conservation NGO—the WildTeam. Villagers possessing knowledge about local culture and forest resource collection processes are selected as VTRT members. Currently, there are 340 VTRT members in 49 teams (7 members per team, with a team leader) working in villages adjacent to the SMF. Key responsibilities of these teams include managing and convincing villagers not to kill stray tigers, saving the villagers and their livestock from tiger attacks, emergency patrolling, transporting tiger victims, and providing emergency medical assistance. Also, VTRTs coordinate with the BFD, local administrative bodies, and other locally influential members to manage an emergency stray tiger situation. VTRT members are connected to villagers with a 24–hour emergency hotline service. In an emergency, VTRT members patrol the area, use firecrackers, manage crowding villagers, and try to chase the tiger back to the forest until the BFD staff rescue the tiger. In 2018, VTRTs helped manage 21 HTC incidents, and as of now, VTRTs helped the BFD to rescue and release 43 tigers back into the forests (Khan et al. 2018).

Inter-Institutional Gaps in Tiger Conservation Governance

Gap Between Formal Constitutional Choice Rules and Informal Constitutional Choice Rules: Legal Pluralism

Despite the presence of strict rules and penalties for tiger killing, about 90–95 percent of local people are unaware of the existing rules (Khan 2004). This knowledge gap is compounded by the extreme poverty and dependency of the local population on forest resources, lack of formal or informal education, inadequate awareness-building programs, and insufficient delivery of services and benefits to local people from the tiger conservation-related projects. On the other hand, the few people who know about the basic features of legal provisions are reluctant to follow them because of weak formal rule enforcement (Khan et al. 2018). The government has not made much effort to alter the local community's hostile attitude toward tigers or to change medication and religious beliefs. Also, most of the conservation rules and policies have mostly been developed based on scientific knowledge and without community consultation. It indicates that unnegotiated attitudes are driving the constitutional-level rules of tiger-conserving formal and informal institutions, which is referred to as legal pluralism in the IIG framework.

*Gap Between Formal Non-Constitutional and Informal Constitutional Choice
Rules: Cultural Mismatch*

The properties of formal non-constitutional and informal constitutional rules indicate
a cultural mismatch between government officials and local community members.
Since the formal constitutional rules are not negotiated with local cultural practices,
government officials have no legal requirement, direction, or capacity to change or
influence local community members' tiger-related beliefs, although they are aware
of these practices. Government officials directly enforce coercive legal instruments
instead of invigorating social services related to tiger conservation, like providing
health care services, creating alternative livelihood opportunities, or incentivizing
tiger protection. Thus, local community members find government officials less
sensitive to community requirements, resulting in conflicting co-existing interactions.
On the contrary, community members have limited access or understanding of the
government officials' legal obligations for conserving tigers. For instance, local
people who inherited wildlife trophies or body parts were asked to inform the local
BFD office within 180 days of enacting the Wildlife (Conservation and Security) Act
2012 for registration to avoid punishment. This notice was published only in a national
newspaper, and no directive was given to local BFD officials to directly inform the
community members. Since local people living around the SMF had limited access
to national newspapers, the majority of them remained unaware of the new act and
failed to register their trophies. This lack of communication suddenly converted many
community members into wildlife criminals, exacerbating conflicting interactions
between government officials and the local community members.

*Gap Between Formal Constitutional Choice Rules and Informal Non-onstitutional
Choice Rules: Institutional Void*

One of the major limitations of existing legal and policy processes is the lack of
community engagement. Most existing community engagement attempts are
the outcomes of non-government conservation agencies (e.g., VTRTs), which
do not receive any incentives from the government. Also, most of the policy and
legal instruments do not sufficiently focus on preventive measures like reducing
community dependence on forest resources, generating alternative and less risky
uses of tiger habitats (e.g., ecotourism), or training local communities about how
to protect lives and resources from a tiger attack without harming or being harmed.
Community members also lack the capacity to interact with policymakers and have
inadequate knowledge about personal and social safety from tiger attacks. Further,
forest users do not have training and equipment (e.g., first aid kits) for immediate
medical support or rapid transportation of a tiger victim to medical facilities. Any
subsequent death enhances local people's negative attitude toward tigers. Due to the
absence of communication with policy makers, local community members cannot
convey their concerns and choices for managing HTC, along with issues related to
easing the delivery of compensation to tiger victim families and public involvement
in rule development and enforcement.

Gap Between Formal Non-Constitutional Choice Rules and Informal Non-Constitutional Choice Rules: Structural Hole

A gap between the non-constitutional choice rules of formal and informal institutions (i.e., structural hole) is evident in mitigating HTCs and in implementing the government's tiger conservation initiatives, resulting mostly from network closure, insufficient knowledge exchange and responsibility sharing between formal and informal institutional actors. The case study demonstrates that the local BFD officials are reluctant to engage in networking with local leaders due to the lack of legal direction, authority, and capacity. Also, the WMNCDK—the BFD's specialized wildlife management unit—has no legal jurisdiction over the forest and faces severe resource and manpower deficits. Wildlife conservation activities are thus done by ad hoc officials coming from other wings of the BFD. These officials consider conservation to be an add-on to their main responsibilities, and they often lack knowledge and logistics for conservation activities and public communication and engagement. Consequently, the officials fail to disseminate information to local community members regarding potential threats and preventive measures.

Voluntary organizations do not have formal status, making it difficult for them to obtain logistic and financial support to maintain their operation. The overall activities and functions of VTRTs gained momentum with the implementation of a USAID-supported Bengal Tiger Conservation Activity (BAGH) project. The project offered training, material support, capacity building, awareness-raising, community engagement, and emergency-response efficiency development services at a larger scale. However, with the completion of the project in 2018, VTRT activities have been reduced significantly. Due to funding limitations, WildTeam activities became infrequent and mostly confined to online-based awareness activities, resulting in less direct contact with local community members, leaders, and VTRT personnel.

Discussion

Our analysis identifies four co-occurring IIGs between the formal and informal institutional regimes of the SMF. A co-existing legal pluralism gap exists between the formal and informal constitutional choice rules (Figure 2a) since most of the formal constitutional rules and policies have been developed without understanding community culture and traditions, and there is an absence of intervention to help change communities' attitudes toward tigers. A cultural mismatch gap exists because government officials are ignorant about local communities' cultural beliefs (e.g., the use of tiger body parts for medicinal and spiritual purposes) (Figure 2b). Also, a lack of effective community engagement for informing policy-making processes about field-level tiger conservation problems is generating a co-existing institutional void (Figure 2c). In addition, network closure and lack of information, resources, and responsibility sharing between the formal institutional actors and local communities have created a co-existing structural hole (Figure 2d). Overall, we obtain four key lessons from this study.

First, legal pluralism has created a parallel journey for formal and informal institutions functioning in tiger conservation. Usually, informal institutional regimes emerge over a historical period to reflect local belief systems and traditional

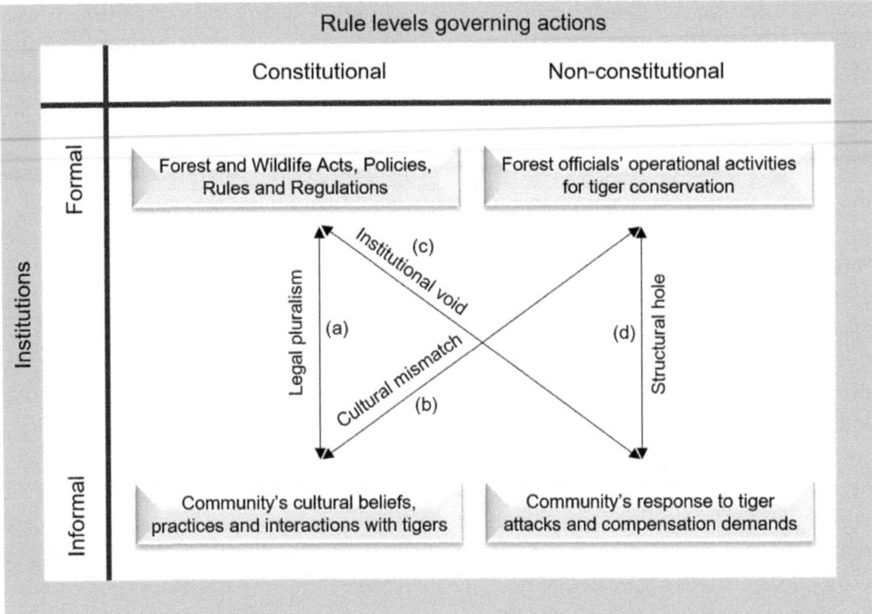

Figure 2: Inter-Institutional Gaps (IIGs) in Conserving Tigers in the Bangladesh Sundarbans. Adapted from Rahman et al. (2017): A framework for analyzing institutional gaps in natural resource governance.

knowledge (Bavinck and Gupta 2014; Rahman et al. 2019). On the contrary, formal institutions may be the outcome of national and international interests and newly emerged scientific knowledge (Bavinck and Gupta 2014). The human settlement and past ruling history of the SMF portrayed tigers as foes and valorized tiger hunters. Social myths and the traditional belief of tiger body parts' healing power incentivized tiger hunting (Hyers 2006). Instead of taking policy interventions to alter community attitudes, ethics, moral arguments, and misconceptions regarding tigers (Novacek 2008; Johansson and Henningsson 2011), the formal constitutional rules of the Bangladesh government took the path of enacting coercive rules to give tigers legal protection, indicating the presence of institutional inertia and lack of institutional innovation in taking account of local tradition and culture (Agrawal 2001). Such disconnects have created competing interactions between the formal and informal institutions, which could be resolved by vigorously studying the cultural dimension of tiger conservation and innovating culturally sensitive policy interventions to help people reimagine their relations with tigers.

Second, building on Acheson (2006) and Hodge (2016), Rahman et al. (2017) conceptualized cultural mismatch as the inadequacy of government officials (i.e., bureaucrats) to engage with community culture and tradition and vice versa, which is prevalent in many post-colonial developing countries. As such, Bangladesh has an incongruity between social norms and bureaucracy (Haque 1997; Zafarullah 2013). Our analysis suggests that operational-level officials do not receive incentives, directives, or training from constitutional-level rules for engaging with local culture, which isolates them as a distinct rule-enforcing group of actors. Moreover, coming from different social, cultural, and economic backgrounds, operational-level bureaucrats cannot comprehend the notion behind certain cultural practices (Rahman et al. 2014). This situation provokes government officials to enforce strict rules indiscriminately, resulting in mistrust and fear among community members toward the government officials. This analysis indicates that enforcing rules is not the only task the government officials can do as a remedy to HTC, rather they need to act as the trusted partners of local people to give protection against tiger attacks.

Third, policymakers need to learn how forest resource users deal with tiger-related challenges in their day-to-day activities to develop practically applicable conservation rules. Community practices are broadly influenced by local leadership and observational knowledge (Hajer 2003; Krishna 2011; Rahman et al. 2014). In this political space, local leadership organizes collective actions in a community to demonstrate their choices to a government (e.g., social movements) or undertake initiatives to implement their collective decisions. In the case of a local community's response to tiger attacks, we observe that the community members are collectively involved in retaliatory killing using strategies that can help them avoid legal actions. Also, community leaders form collective organizations supported by non-government organizations to protect communities protection against tiger attacks, which have shown much success and social acceptance. However, the Bangladesh government has shown less interest in recognizing and supporting these organizations. For instance, the government has abandoned effective community services such as ferrying tiger victims from the forest to medical centers. Since policymakers do not maintain direct communication with community members, the decision-making processes always lack empirical connectivity.

Finally, a structural hole emerges when cohesiveness is high among actors functioning in a resource system (Burt 2000, 2004; Gargiulo and Benassi 2000). Strong cohesiveness may reinforce group belief, knowledge, and homogenous behavior and abstain from information sharing, collective interventions, and coordination among actors. Intermediary actors (e.g., NGOs), having resources and brokering capacity, can effectively bridge the functional disconnects among the actor groups (Gargiulo and Benassi 2000; Burt 2004). Our analysis suggests that the bureaucratic system of Bangladesh is not only incongruent with social norms and culture but also with everyday social practices. While the tiger conservation activities could have been implemented based on coordinated interventions between operational-level government officials and community members, the government has done little to mobilize specialized agencies like WMNCDK to build coordination. Non-government agencies like the WildTeam can play innovative roles by allocating expertise and resources to organize community members for mediating the structural gap between the operational level government officials and local community members. However, we observe that this gap has resurfaced with replenishing the agency's resources.

Conclusion

Although government policies and socio-cultural factors simultaneously influence tiger conservation efforts in Bangladesh, their interactions are poorly understood. These interactions are important to help develop coordination between policies and social practices in tiger conservation efforts. Using an institutional perspective and applying the IIG framework, this study identifies coordination gaps at different levels of the Bangladesh government's formal institution and tiger-affected communities' social norm-driven informal institutions. These gaps significantly reduce the efficiency of government policy to conserve tigers, as the government heavily relies on enforcing its legal instruments rather than developing culturally and socially sensitive actions.

In summary, the study finds that government policies could not mobilize any action to change social myths and beliefs contributing to tiger killing. Also, the lack of policy direction and resource allocation to local-level government officials to act in culturally sensitive ways has created distrust between government officials and local community members. Also, policy-making processes lack mechanisms to learn from the forest resource users' day-to-day activities and lived experiences with tiger conservation challenges. As such, policy-making processes lack connectivity with empirical situations. Finally, there is a lack of coordinated functioning of local-level government actors and resource users to conserve tigers. The government officials do not have adequate resources, authority, and expertise to build fruitful communication with local communities. While non-government agencies have been coordinating the functions of government officials and resource users, they do not receive much policy support. As a result, the community members do not receive sufficient government services to avoid tiger attacks. All these factors contribute to the local community's negative attitudes toward tigers.

Based on this study, we recommend that the government conduct an in-depth study on the sociocultural aspects of tiger conservation to help develop culturally sensitive policies and intervene to alter cultural attitudes and beliefs. These interventions are socially acceptable if local-level government officials get policy directives to learn local traditional and cultural practices associated with tigers and act to provide alternative services (e.g., health care services, security, and protection). Also, policies should not be developed in the absence of direct communication or input from local resource users, who are the most frequent tiger victims. Intermediary agencies must also be supported with resources to facilitate coordination between government officials and local resource users. We conclude that conservation policies and actions would be effective only when they are culturally engaged, more participatory, and problem-oriented.

Applying the IIG framework, this chapter shows that conservation policy-making needs to be compatible with local culture and tradition. Community participation should not be considered only as a process of involving community members in governments' biodiversity conservation efforts; rather, it should thoroughly investigate and learn the conservation culture of a society and develop strategies for dealing with the social challenges of conservation. Cultural learning should be exchanged with operational-level government officials to improve their understanding of avoiding conflicts with community members.

References

Acheson, J.M. 2006. Institutional failure in resource management. Annual Review of Anthropology 35: 117–134.

Agrawal, A. 2001. Common property institutions and sustainable governance of resources. World Development 29: 1649–1672.

Ahmad, I.U., C.J. Greenwood, A.C.D. Barlow, M.A. Islam, A.N.M. Hossain, M.M.H. Khan et al. 2009. Bangladesh tiger action plan 2009–2017. Bangladesh Forest Department, Ministry of Environment, Forest and Climate Change, Dhaka, Bangladesh.

Aziz, A., A.C.D. Barlow, C.C. Greenwood and A. Islam. 2013. Prioritizing threats to improve conservation strategy for the tiger Panthera tigris in the Sundarbans Reserve Forest of Bangladesh. Oryx 47: 510–518.

Aziz, A. and A. Paul. 2015. Bangladesh Sundarbans: Present Status of the Environment and Biota. Diversity 7: 242–269.

Aziz, M.A. 2017. Population status, threats, and evolutionary conservation genetics of Bengal tigers in the Sundarbans of Bangladesh. University of Kent.

Barlow, A.C.D. 2009. The Sundarbans tiger: Adaptation, population status and conflict management. University of Minnesota.

Barlow, A.C.D., M.I.U. Ahmed, M.M. Rahman, A. Howlader, A.C. Smith and J.L.D. Smith. 2008. Linking monitoring and intervention for improved management of tigers in the Sundarbans of Bangladesh. Biological Conservation 141: 2032–2040.

Bavinck, M. and J. Gupta. 2014. Legal pluralism in aquatic regimes: a challenge for governance. Current Opinion in Environmental Sustainability 11: 78–85.

BFD. 2010. Integrated Resources Management Plans for the Sundarbans: 2010–2020. Dhaka, Bangladesh.

BFD. 2016a. National Tiger Recovery Program of Bangladesh 2017-2022. Dhaka, Bangladesh.

BFD. 2016b. Bangladesh Wildlife Conservation Master Plan: 2015–2035. Bangladesh Forest Department, Government of the People's Republic of Bangladesh., Dhaka, Bangladesh.

BFD. 2019. The Sundarbans at a glimpse. Office of the Conservator of Forests, Khulna, Khulna.

BFD. 2021. International convention, treaties and the position of Bangladesh. Bangladesh Forest Department, Government of the People's Republic of Bangladesh. http://www.bforest.gov.bd/site/page/b81a2e19-f10f-480c-9a88-b5e572197f39/International-conventions.

Broad, S. and R. Damania. 2010. Competing Demands: Understanding and Addressing the Socio-economic Forces that Work for and against Tiger Conservation. Page October. Washington, DC.

Burt, R.S. 2000. The Network Structure Of Social Capital. Research in Organizational Behavior 22: 345–423.

Burt, R.S. 2004. Structural holes and good ideas. American Journal of Sociology 110: 349–399.

Chakrabarti, R. 2009. Local People and the Global Tiger: An Environmental History of the Sundarbans. Global Environment 3: 72–95.

Danielsen, F., M.K. Sørensen, M.F. Olwig, V. Selvam, F. Parish, N.D. Burgess et al. 2005. The Asian tsunami: a protective role for coastal vegetation. Science (New York, N.Y.) 310: 643.

Department of Environment. 2016. National Biodiversity Strategy and Action Plan (NBSAP) of Bangladesh 2016–2021. Department of Environment, Ministry of Environment, Forests & Climate Change, Government of the People's Republic of Bangladesh, Dhaka, Bangladesh.

Elo, S. and H. Kyngäs. 2008. The qualitative content analysis process. Journal of Advanced Nursing 62: 107–115.

Gargiulo, M. and M. Benassi. 2000. Trapped in Your Own Net? Network Cohesion, Structural Holes, and the Adaptation of Social Capital. Organization Science 11: 183–196.

GoB. 1927. The Forest Act, 1927. Page 43. Bangladesh.

GoB. 1995. The Bangladesh Environment Conservation Act 1995. Pages 153–166 Bangladesh Gazette.

GoB. 2012. WILDLIFE (CONSERVATION AND SECURITY) ACT, 2012. Bangladesh.

GoB. 2017. Protected Area Management Rules 2017. Page 16. Bangladesh.

Gopal, B. and M. Chauhan. 2006. Biodiversity and its conservation in the Sundarban Mangrove Ecosystem. Aquatic Sciences 68: 338–354.

Hajer, M. 2003. Policy without polity? Policy analysis and the institutional void. Policy Sciences 36: 175–195.

Haque, M.S. 1997. Incongruity Between Bureaucracy and Society in Developing Nations: A Critique. Peace & Change 22: 432–462.

Hardy, S.D. and T.M. Koontz. 2009. Rules for collaboration: institutional analysis of group membership and levels of action in watershed partnerships. Policy Studies Journal 37: 393–414.

Hodge, J.M. 2016. Colonial Experts, Developmental and Environmental Doctrines, and the Legacies of Late British Colonialism. Page in K.O. Christina Folke Ax, Niels Brimnes, Niklas Thode Jensen, editor. Cultivating the Colonies. Ohio University Press, Ohio.

Hyers, L.L. 2006. Myths used to legitimize the exploitation of animals: An application of Social Dominance Theory. Anthrozoös 19: 194–210.

Inskip, C., M. Ridout, Z. Fahad, R. Tully, A. Barlow, C.G. Barlow et al. 2013. Human–Tiger Conflict in Context: Risks to Lives and Livelihoods in the Bangladesh Sundarbans. Human Ecology 41: 169–186.

Islam, S., M. Rahman and S. Chakma. 2014. Plant diversity and forest structure of the three protected Areas (Wildlife Sanctuaries) of Bangladesh Sundarbans: Current Status and Management Strategies. pp. 127–152. *In*: Latiff, A., I. Faridah-Hanum, M. Ozturk and Khalid Rehman Hakeem (eds.). Mangrove Ecosystems of Asia. Springer, New York.

Johansson, M. and M. Henningsson. 2011. Social-psychological factors in public support for local biodiversity conservation. Society and Natural Resources 24: 717–733.

Johnson, M., K. Karanth and E. Weinthal. 2018. Compensation as a Policy for Mitigating Human-wildlife Conflict Around Four Protected Areas in Rajasthan, India. Conservation and Society 16: 305–319.

Khan, M.M. 2004. Ecology and conservation of the Bengal tiger in the Sundarbans mangrove forest of Bangladesh. University of Cambridge.

Khan, M.M.H., M.M. Ahsan, Y.V. Jhala, Z.U. Ahmed, A.R. Paul, M.J. Kabir et al. 2018. Bangladesh Tiger Action Plan, 2018-2027. Bangladesh Forest Department, Ministry of Environment, Forest and Climate Change, Dhaka.

Krishna, A. 2011. Gaining access to public services and the democratic state in India: institutions in the middle. Studies in Comparative International Development 46: 98–117.

McGinnis, M.D. and E. Ostrom. 2014. Social-ecological system framework: initial changes and continuing challenges. Ecology and Society 19(2): 30. http://dx.doi.org/10.5751/ES-06387-190230.

MoEF. 2016. Bangladesh National Conservation Strategy (2016–2031). Dhaka, Bangladesh.

Mondol, S., K.U. Karanth and U. Ramakrishnan. 2009. Why the Indian Subcontinent Holds the Key to Global Tiger Recovery. PLoS Genetics 5:e1000585.

North, D.C. 1991. Institutions. Journal of Economic Perspectives 5: 97–112.

Novacek, M.J. 2008. In the Light of Evolution. Page In the Light of Evolution. National Academies Press, Washington, D.C.

Ostrom, E. 2005. Understanding institutional diversity. Page New Jersey. Princeton University press, Princeton, NJ.

Ostrom, E. 2007. A diagnostic approach for going beyond panaceas. Proceedings of the National Academy of Sciences 104: 15181–15187.

Ostrom, E. 2009. A General Framework for Analyzing Sustainability of Social-Ecological Systems. Science 325: 419–422.

Ostrom, E., R. Gardner and J. Walker. 1994. Rules, Games, and Common-Pool Resources. Page Ann Arbor, MI: University of Michigan Press. https://doi. org/10.3998/mpub.9739. University of Michigan Press, Michigan, USA.

Rahman, H.M.T., J.Y.T. Po, A.S. Saint Ville, N.D. Brunet, S.M. Clare, S. Darling et al. 2019. Legitimacy of different knowledge types in natural resource governance and their functions in inter-institutional gaps. Society & Natural Resources 32: 1344–1363.

Rahman, H.M.T., S.K. Sarker, G.M. Hickey, M. Mohasinul Haque and N. Das. 2014. Informal Institutional Responses to Government Interventions: Lessons from Madhupur National Park, Bangladesh. Environmental Management 54: 1175–1189.

Rahman, H.M.T., A.S. Saint Ville, A.M. Song, J.Y.T. Po, E. Berthet, J.R. Brammer et al. 2017. A framework for analyzing institutional gaps in natural resource governance. International Journal of the Commons 11: 823–853.

Rastogi, A., G.M. Hickey, R. Badola and S.A. Hussain. 2014. Understanding the Local Socio-political Processes Affecting Conservation Management Outcomes in Corbett Tiger Reserve, India. Environmental Management 53: 913–929.

Rasul, G. 2007. Political ecology of the degradation of forest commons in the Chittagong Hill Tracts of Bangladesh. Environmental Conservation 34: 153–163.

Richardson, S., A.C. Mill, D. Davis, D. Jam and A.I. Ward. 2020. A systematic review of adaptive wildlife management for the control of invasive, non-native mammals, and other human–wildlife conflicts. Mammal Review 50: 147–156.

Roy, A.K.D., K. Alam and J. Gow. 2013. Community perceptions of state forest ownership and management: A case study of the Sundarbans Mangrove Forest in Bangladesh. Journal of Environmental Management 117: 141–149.

Saif, S., H.M.T. Rahman and D.C. MacMillan. 2018. Who is killing the tiger Panthera tigris and why? Oryx 52: 46–54.

Saif, S., A.M. Russell, S.I. Nodie, C. Inskip, P. Lahann, A. Barlow et al. 2016. Local usage of Tiger Parts and Its Role in Tiger Killing in the Bangladesh Sundarbans. Human Dimensions of Wildlife 21: 95–110.

Sandilyan, S. and K. Kathiresan. 2012. Mangrove conservation: a global perspective. Biodiversity and Conservation 21: 3523–3542.

Sarker, S.K., J. Matthiopoulos, S.N. Mitchell, Z.U. Ahmed, M.B. Al Mamun and R. Reeve. 2019a. 1980s–2010s: The world's largest mangrove ecosystem is becoming homogenous. Biological Conservation 236: 79–91.

Sarker, S.K., R. Reeve and J. Matthiopoulos. 2021. Solving the fourth-corner problem: forecasting ecosystem primary production from spatial multispecies trait-based models. Ecological Monographs 91:ecm.1454.

Sarker, S.K., R. Reeve, N.K. Paul and J. Matthiopoulos. 2019b. Modelling spatial biodiversity in the world's largest mangrove ecosystem-The Bangladesh Sundarbans: A baseline for conservation. Diversity and Distributions 25: 1–14.

Sarker, S.K., R. Reeve, J. Thompson, N.K. Paul and J. Matthiopoulos. 2016. Are we failing to protect threatened mangroves in the Sundarbans world heritage ecosystem? Scientific Reports 6: 21234.

Siddiqi, N.A. 2001. Mangrove forestry in Bangladesh. Institute of Forestry and Environmental Sciences, University of Chittagong, Chittagong, Bangladesh.

Walston, J., J.G. Robinson, E.L. Bennett, U. Breitenmoser, G.A.B. da Fonseca, J. Goodrich et al. 2010. Bringing the Tiger Back from the Brink—The Six Percent Solution. PLoS Biology 8: e1000485.

Wilting, A., A. Courtiol, P. Christiansen, J. Niedballa, A.K. Scharf, L. Orlando et al. 2015. Planning tiger recovery: Understanding intraspecific variation for effective conservation. Science Advances 1:e1400175.

Zafarullah, H. 2013. Bureaucratic culture and the social-political connection: the bangladesh example. International Journal of Public Administration 36: 932–939.

5

Applying an Access Lens to Understand Equity in a Polycentric Governance Regime

Why Rights Alone May be Insufficient to Advance Indigenous Fishery Development

Hekia Bodwitch,[1,]* *Megan Bailey*[2] and *John Reid*[3]

Introduction

The complexity and magnitude of contemporary socio-environmental problems has given rise to claims that centralized, state-led environmental governance regimes are inadequate and, in many cases, unable to adapt to change and uncertainty (Ostrom 2010, 2012). Institutionally diverse, polycentric governance regimes, which are comprised of multiple autonomous decision-making units, have been presented as an alternative (Ostrom et al. 1961). Theoretically, the option for citizens to organize to address additional needs through diversity and polycentricity may provide the ability to mitigate risk due to the reduced likelihood that all decision-making units' responses will fail (Ostrom 2012; Carlisle and Gruby 2019). Polycentric regimes are also understood to hold greater adaptive capacities if options for learning and

[1] University of Alaska Southeast, School of Arts and Sciences, 11066 Auke Lake Way, Juneau, AK 99801 USA.

[2] Dalhousie University Faculty of Science, Marine Affairs Program, 1355 Oxford Street, Rm 805, 8th Floor Life Sciences Centre (Biology), Halifax, Nova Scotia, Canada B3H 4R2.
Email: Megan.Bailey@dal.ca

[3] Ngāi Tahu Research Centre, University of Canterbury, Private Bag 4800, Christchurch 8140, Aotearoa New Zealand.
Email: John.Reid@canterbury.ac.nz

* Corresponding author: hebodwitch@alaska.edu

collaboration exist between decision-making units (Aligica 2014). In the field of natural resource governance, the concept gained considerable momentum following E. Ostrom's promotion of polycentricity as a means to address climate change governance challenges (Ostrom 2009, 2010), so much so that advocates of polycentric regimes have been accused of adopting a panacea mindset—advocating for an approach without accounting for its efficacy (Young et al. 2018; Carlisle and Gruby 2019).

More recently, scholars have examined the specific mechanisms through which polycentric governance regimes lead to particular outcomes. Key elements identified include an overarching set of rules that allows citizens to organize into decision-making units that possess "some degree of autonomy" in order to address their needs (Aligica 2014; Carlisle and Gruby 2019). In studies of natural resource management, however, analyses describing the criterion necessary to advance polycentric governance initiatives have primarily focused on the capacity of governance systems to respond to ecological changes (Pattberg and Widerberg 2016; Morrison 2017; Steffen et al. 2018; Carlisle and Gruby 2018; Acton et al. 2021; Urlich et al. 2022). Despite a now widespread understanding that environmental issues are also social justice issues (Mohai et al. 2009), few studies of natural resource governance have examined relationships between institutional diversity, polycentric governance, and equity outcomes.[1] In this chapter, we discuss and illustrate a method for examining these relationships.

Our approach draws on access theory (Ribot and Peluso 2003) and is based on the premise that processes beyond rights alone are required to access or derive benefit from resources. Instead, an individual's ability to access resources can be affected by institutions in the form of rights, rules, or norms (i.e., Rahman et al. 2017) as well as by various ecological, social, and economic dynamics. As Ribot and Peluso (2003) highlighted, these dynamics include identity politics, influenced by discursive representations of peoples and places, and material conditions, such as biophysical processes and access to knowledge or capital. They change over time and are contingent on an individual's geographical location as well as the actions of others (Ribot and Peluso 2003; Peluso and Ribot 2020).

These dynamics, we emphasize, influence which institutions come to have the strongest effects on the ability of individuals (or groups) to benefit from resources. These institutions manifest collectively as "networks" or "bundles" of power (Ribot and Peluso 2003) that establish social hierarchies and explain why some individuals remain excluded while others gain and maintain privileged access to particular resources (Peluso and Ribot 2020). Given that multiple institutions can affect resource use, we use access analysis to identify which aspects of institutionally diverse contexts warrant reworking to support equity outcomes. Drawing on environmental justice scholarship (e.g., Holifield et al. 2018), we conceive of equity as affected by the distributions of burdens and benefits from environmental resources.

---

[1] For exceptions, see De Wit and Mourato (2022) and Diver et al. (2022).

In our conception, access analyses involve the identification of (1) conditions that must be met for an individual or group to benefit from a particular resource, which, in turn, affect how burdens and benefits from environmental resources are distributed; (2) dynamics that constrain an individual's or group's ability to meet these conditions; (3) institutions that mediate the inequitable effects of dynamics identified in phase 2.

This form[2] of access analysis requires one to account for a resource user's position in relation to other resource users and the governance processes that give rise to this positioning. As a result, our access analysis is multi-scalar and multi-sited. To conduct this analysis, we align with feminist science studies scholars and start from the standpoints of subjugated individuals (Haraway 1988; Harding 2004; Bodwitch 2014). Not because, as Haraway (1998) argues, these standpoints are necessarily more "innocent" but rather because individuals from historically subjugated groups are likely to know how multiple institutions affect equity. As evidenced by their survival, there is a decent chance these individuals are likely to know both the dominant, state-authorized institutions governing resource access, as well as the strategies individuals have engaged to work around limitations of dominant systems.

In what follows, we illustrate how access analyses can account for relationships between institutional diversity, polycentricity, and equity outcomes by examining an Indigenous rights reconciliation initiative in New Zealand. Questions of how polycentric governance regimes can advance equity are especially significant in contexts where Indigenous groups hold political or legal authority to advance self-governance regimes. In United Nations (UN) member states, such political authorities have been imparted through ratification of the 2007 UN Declaration on the Rights of Indigenous Peoples and can hold legal traction when Indigenous groups also have treaty or aboriginal title rights, as is the case in many former British colonies (Engle 2010; Parsons et al. 2021).

In recognizing rights, nation-state governments can transfer or share governance responsibilities by establishing polycentric regimes. In these regimes, Indigenous autonomy is never absolute, given the claims to territorial sovereignty that nation-state governments maintain. New Zealand is an important case to examine how Indigenous-state polycentric governance arrangements can support equity, given the international acclaim the nation has received for its efforts to reconcile Māori grievances (Anaya 2011). However, in New Zealand, the extent to which the resulting polycentric governance arrangement has mitigated inequitable distributions of burdens and benefits from resources is called into question when one examines the nation's fisheries governance through an access lens.

Background: Polycentric Governance and Indigenous Rights in New Zealand's Fisheries

In New Zealand, Māori obtained commercial fishing rights through a series of agreements with the New Zealand government between 1989 and 2004 (Table 1).

---

[2] We emphasize this as a unique form of access analysis in light of the proliferation of scholarship that deploys access analyses using various approaches (Myers and Hanson 2020).

These rights came in the form of an Individual Transferable Quota (ITQ) to Māori collectively to govern as they saw fit. In fisheries managed by ITQ systems, governments set a limit on the total amount of fish that can be taken for commercial sale and allocate to commercial fishers a percentage of the cap, called a quota (Sumaila 2010). One goal of these systems is to reduce overcapacity in fisheries, or essentially, the number of fishers on the water. This is done by making it possible for fishers to sell or lease their quota, a provision that serves to consolidate quota and privatize fisheries (Carothers and Chambers 2012). In 1986, the New Zealand government implemented one of the world's first comprehensive Individual Transferable Quota systems, previously conceptualized as a theoretical model (Campbell 1984; Copes 1986; Christy 1973). However, shortly after system implementation, in response to a Māori claim, New Zealand courts ruled that the system violated Māori Treaty rights because the government had to assume it owned the nation's fisheries to privatize them (Boast 1999). Māori Treaty rights are protected by the Treaty of Waitangi, a document signed in 1840 between Māori Iwi (tribes) and the British government, which made New Zealand a colony of Britain and promised to Māori "full exclusive and undisturbed possession" of their lands, fisheries, and other valued relations unless Māori agreed to relinquish them to the British (Orange 2004).

ITQ systems have been implemented worldwide, with criticism due to the exclusionary effects of this market-based regime for resource management (Pinkerton and Davis 2015). In New Zealand, the government's system for ITQ rights allocations led to exclusions that corresponded to racialized categories of difference. The New Zealand government initially allocated quota rights based on fishers' reported catch histories and then made the rights tradable as a marketized commodity (Sissenwine and Mace 1992). This policy disproportionately affected Māori, who were unlikely to have reported catch histories to obtain quota initially or capital necessary to purchase quota after allocation (Bodwitch 2017a). Many Māori fishers were fishing off Māori reserved land in areas where the government had not developed ports or offered subsidies to fishers (Waitangi Tribunal 1992). In interviews, fishers fishing in these locations reported that they did not understand the potential benefits of reporting their catches to the government, and they viewed these fisheries as Māori-owned (Bodwitch 2017b).

Following a 1989 interim agreement, the 1992 Fisheries Settlement was designed partly to mitigate the contemporary impacts of Māori fishers' exclusion from quota system implementation, most of whom were fishing close to shore. It also promised to absolve grievances related to processes of exclusion resulting from colonial policies that restricted Māori groups' access to capital necessary to fish species further offshore. The 1989 interim agreement granted to Māori collectively 10 percent of the quota for the 26 fish species that were already in the system at that time. In the 1992 Settlement, Māori also obtained 20 percent of the quota for species to be added later (Boast 1999), shares in the nation's largest fishing company, as well as limited authorities to govern customarily significant fisheries and the ability to harvest fish for non-commercial use above daily recreational catch limits (McCormack 2010). In allocating quota rights to Māori, the 1989 and 1992 agreements created a polycentric governing system for commercial fishing rights allocations in New Zealand, in which Indigenous groups held limited autonomy to govern how Settlement quota might be

Table1: Contemporary Māori fishing rights: key legislation.

Treaty of Waitangi 1840	• Established New Zealand as a colony of Britain; • Promised Māori "full exclusive and undisturbed possession" of their fisheries.
Fisheries Amendment Act 1986	• Established an ITQ system for New Zealand's fisheries.
NZ Māori Council v. Attorney General 1987	• The judge's ruling halted the New Zealand government's quota allocation initiatives until a settlement was reached with Māori.
Māori Fisheries Act 1989	• The interim settlement between Māori and the New Zealand government that allowed New Zealand's ITQ system to proceed; • Transferred 10% of available quota to the newly established Māori Fisheries Commission (MFC); • MFC was tasked to establish Aotearoa Fisheries Limited (AFL) to hold 50% of the settlement quota and lease the remainder.
Treaty of Waitangi (Fisheries Claims) Settlement Act 1992	• Set up customary fishing rights as non-commercial rights; • Gave MFC (reconstituted as Treaty of Waitangi Fisheries Commission, ToWFC) NZD 150M to go into a joint venture with Sealords, owner of 22% of available quota; • Permitted ToWFC to allocate Settlement assets; • Allocated 20% of the new quota for stocks not in ITQs to ToWFC.
Māori Fisheries Act 2004	• Disbanded the ToWFC, allocated its quota and financial assets to Aotearoa Fisheries Limited (AFL), established Te Ohu Kaimoana (TOKM) as the sole voting shareholder; • Tasked TOKM with a responsibility to allocate quota to Mandated Iwi Organizations (MIOs); • Established that MIOs would be allocated quota for deep-sea species (75% fished 300m from shore) based on population (75%) and coastline (25%) and inshore species based primarily on coastline.

allocated. Within this system, the government retained authority to determine how non-Settlement quota rights would be allocated, which was done through the ITQ system's market-based mechanism (Boast 1999).

Māori fishers negotiating the claim believed the Settlement would give them the ability to support their livelihoods through fishing. For these fishers, cash-based fish sales enabled their access to boats and gear necessary for fishing prior to the establishment of the ITQ system. The ITQ system made these sales illegal. However, since the Settlement, few Māori have been fishing (Rout et al. 2019). Most Māori-owned quota is leased to non-Indigenous fishing operations. Here, we use an access analysis to examine why this is the case. Drawing on Māori fishers' experiences, we identify four key conditions that affect a fisher's ability to access and obtain benefits from fisheries. These conditions include access to fishing rights, means of production, markets, and fish. We find that New Zealand's Indigenous rights recognition initiative has imparted the means for Māori to augment some, but not all, barriers to access associated with the diverse institutional landscape that governs how Māori fishers can meet these conditions. In what follows, we first describe the methods for collecting the data that informs this access analysis. We then describe the various dynamics and institutions that affect Māori fishers' abilities to access rights, means of production, markets, and fish. In the concluding discussion, we explore the implications of this access analysis for advancing equity in the governance of natural resources in New Zealand and elsewhere.

Data Collection Methods

Our study draws on the experiences of Māori fishers and leaders from the Ngāi Tahu Iwi, or tribe, based in New Zealand's South Island and one of the largest and wealthiest Iwi in New Zealand (Figure 1). Ngāi Tahu Indigenous fishery development initiatives represent one of the longest-running efforts of a Māori group to use quota rights to support Māori fishers (Bodwitch 2017a). The data analyzed was derived from interviews, focus groups, and participant observation conducted by the lead author from 2015 to 2019. Approximately 150 interviews and 40 focus groups with Māori and non-Māori fishers, scientists, and leaders were conducted, many of which were follow-up interviews, and many were unrecorded given the sensitive nature of the topics discussed. All interviews and focus groups were semi-structured and explored barriers Māori and non-Māori leaders faced and strategies they deployed in their attempts to develop fishing operations. The lead author also engaged in processes of participant observation by working as a fisher on a Māori fisher's boat and living in a Māori fisher's hut in a coastal community on New Zealand's South Island. Additionally, she has held a position with a Ngāi Tahu research institute. Bodwitch grew up in a rural region of upstate New York. She is non-Indigenous and has family connections to fishing and farming communities in New Zealand (for more on methods, see Bodwitch 2017b; Bodwitch et al. 2022). This study's focus on fisheries was derived from conversations between Bodwitch and Māori community members in 2010 and 2011, who described the socioeconomic challenges Māori coastal communities faced following fishers' exclusion in the wake of ITQ system implementation.

Results: Institutional Diversity and the Governance of Māori Fishers' Access

Access to Rights

The Settlement granted access to rights, but how it did so corresponded to a lengthy deliberation process, such that by the time Māori obtained fishing rights, the contexts they were fishing in had changed. Specifically, the industry had experienced consolidation as processors purchased quotas from fishers, resulting in a reduction of the size of the fishing fleet (a stated goal of ITQ system implementation), as well as increased processor control over market access. Processors who owned quota outcompeted non-quota-owning processors, and many processors shut down, reducing market access options for fishers. Processors had an advantage over fishers for quota purchases, as they could leverage land-based infrastructure for bank loans (Memon and Cullen 1992), while fishers could not leverage boats for loans. Processors were incentivized to purchase quotas to ensure they would have fish to process. Transferring quota to a Māori trust in the Settlement also facilitated consolidation by creating an entity with a quota to lease within a regulatory regime in which quota leases enabled processors (and others) to overcome consolidation limits. Consolidation limits, as per the 1996 Fisheries Act, were already relatively high at 35 percent for most species, with possibilities for government-granted exemptions (Yandle and Dewees 2008).

As noted, the 1992 Settlement held a dual imperative to account for intragenerational and intergenerational equity. The agreement sought to address contemporary Māori fishers' exclusion due to the ITQ system's allocation provisions *and* mitigate the effects of historical processes that restricted all Māori individuals' abilities to fund fishing operations. The choice facing Iwi leaders has thus been whether to allocate Settlement quota to fishers, which means the Iwi forfeit the potential lucrativeness of the quota asset as capital, or to lease the asset and use lease profits to purchase more quota and invest in socio-economic development.

Quota leasing was the primary Māori quota management strategy from 1992–2004, during which time a Māori trust managed the Settlement quota while Māori groups worked out how to divide the assets between them. In 2004, Māori leaders decided to split the Settlement quota based on population and coastline (McCormack 2021). This decision prompted, in some cases, extensive processes for determining Iwi coastal boundaries and resulted in a situation where the vast majority of Iwi did not receive their Settlement quota package for more than a decade after the 2004 agreement. Upon receiving their quota, most Māori groups continued to lease it almost exclusively to non-Māori operations, in part to ensure that the benefits of this one-off Settlement persisted for future generations. Māori-owned quota leasing practices have proved lucrative to an extent. As of 2019, Māori owned over 30 percent of the nation's fishing quota. Most is fished by non-Māori (Rout et al. 2019) and thus, while these access rights promote the attainment of economic benefits through royalties, they have not led to an equal attainment of fisheries development. Māori fishers are rarely able to outcompete larger operations for quota lease agreements. Quota leasing to larger operations is a particularly attractive option for Māori groups with smaller quota packages (i.e., those with smaller coastlines and populations), whose quota amount for certain species may be insufficient to cover the costs of fishing operations, especially if these species are located offshore and are typically fished with expensive equipment.

The structural imperative for quota to serve two purposes, address historical as well as contemporary processes of exclusion, might be offset if Māori fishing operations were more lucrative and could pay higher quota lease prices than non-Māori operations. Why they are not is linked strongly to the fact that a diversity of institutions, not only governance of quota rights allocations, affect possibilities for fishery development in New Zealand. These institutions include rules governing how fishing may occur and, in turn, the means of production necessary to fish and how fish can be traded or sold and how other resource users, including farmers upstream, can engage with resources that affect fish habitat. While the allocation of fishing rights is overseen by a polycentric, Indigenous-state governance regime, the additional institutions affecting fishery development are governed by a monocentric, state-led regulatory regime. Although some Indigenous fishers are developing strategies to work within these systems, these institutions continue to exclude the majority of Indigenous fishers and, in certain instances, as is the case in fisheries depleted by upstream land use (Bodwitch et al. 2022, 2024), affect processes of exclusion for all Indigenous fishers.

Access to Means of Production

The financial costs associated with becoming a fisher include purchases like a boat, gear, insurance, and fuel, as well as labor costs and time costs associated with accessing knowledge and learning new technologies. The costs associated with accessing the means of production necessary to fish are determined partly by state regulations around how fishing practices may occur (e.g., boat safety requirements and labor laws). The costs associated with such regulations are exacerbated by extensive reporting requirements that increase the "administrative burdens" fishers must overcome to legally fish (Moynihan et al. 2014; Bodwitch et al. 2021). These criteria disproportionately impact Māori due to the time delay between ITQ system implementation in 1986 and 2004 or later, when Iwi obtained their Settlement quota (Table 1). Māori who were fishing at the time of ITQ implementation could not keep up their gear because they could not access capital and thus saw their equipment fall into disrepair. In order to obtain access to the means of production necessary to support fishing operations, fishers have engaged in processes of self-exploitation (i.e., working long hours after fishing to repair boats or gear), relied on income obtained from non-fishing jobs, and called upon various forms of social capital (e.g., borrowing neighbor's nets). In certain contexts, Iwi have provided support to fishers, made possible in part through the transfer of financial assets to Māori groups in Settlements. When doing so, an Iwi effectively subsidizes one or two individual fishers through the divestment of funds that might otherwise be used to support other Iwi members or invest for future generations. Nonetheless, these subsidies could be read as evidence of the polycentric system addressing the problem it is designed to, to the extent profits from quota leasing facilitate Māori fishery development. However, Māori fishers who have developed operations through Iwi support have faced additional cost-related challenges due to state-level regulations around market access as well as the New Zealand government's support for intensified forms of agricultural production upstream.

Access to Markets

As in other commercial ventures, Māori fishers can offset the cost of their fishing operation by selling their fish and navigating markets. While this is true in theory, in practice, the consolidation of the processing sector after the ITQ systems establishment and after the 1992 Fisheries Settlement has limited Māori fishers' market options. Consolidation of the processing sector in New Zealand reflects the government's requirement that all fish caught for commercial sale be sold through licensed fish receivers, such that fishers cannot sell directly off the boat (Bodwitch 2017a). It is also a reflection of the quota accumulation practices that certain, mainly non-Māori, processors engaged in following the implementation of the ITQ system.

Immediately after the ITQ system's implementation, processors bought quotas to ensure their access to fish. Processors had an advantage over fishers in obtaining cash to purchase quota because, unlike fishing boats, their physical investments in land-based infrastructures could be leveraged for loans. Māori were unlikely to be processors due to the legacies of colonial policies that restricted Māori access to capital (Waitangi Tribunal 1992). Even when Iwi own their own processing companies, they remain in competition with Iwi fishers for quota, given that the Iwi processor, as is the case with all processors, is reliant on fishers landing their fish to the plant, creating an incentive for the Iwi processor to outbid Iwi fishers to control quota (Bodwitch 2017a). The 1992 Settlement's transfer of rights, on its own, has not destabilized the power processors hold nor resulting social hierarchies. These dynamics are illustrated in the case of Māori fishers Roger and Paul Tainui (the Tainuis).

The Tainuis were excluded from the New Zealand government's initial quota allocation processes due to a lack of knowledge of the benefits of reporting catches. The Tainuis later obtained commercial fishing rights from the Ngāi Tahu Iwi to fish for eel, a historically important species for Māori, deemed by Iwi leaders to hold too great a cultural significance to lease to non-Māori. When the Tainuis began fishing Settlement quota, less than a handful of eel processors were operating on the South Island. To improve their options for market access, they sought to vertically integrate by developing their processing plant. Their ability to do this was contingent on their ownership of land, which, in their case, was a parcel of family-owned land located an hour's drive from their fishing site, Te Waihora/Lake Ellesmere, a brackish lagoon on the South Island's east coast (Figure 1). Māori fishers have a long history of processing eel and other species on Te Waihora's shores and continue to do so for fish caught on customary or recreational permits. However, traditional Māori fish processing occurs in unenclosed, outdoor spaces, which are ineligible for licensure under New Zealand's processor compliance standards. Requirements for processing indoors pose challenges for Māori fishery development, not least because indoor facilities require land to establish operations upon. Many Māori have access to land, but often, this land is ineligible for investment, a reflection of colonial regulations that prohibited the sale of Māori-owned land to non-Māori (Waitangi Tribunal 2016). These regulations were designed to mitigate the exclusion of Māori from landownership following land purchases by colonial actors but have led to a situation where much Māori-owned land is owned by upwards of 300 people and is considered uneconomic (Rout et al. 2020). The Tainuis have access to a plot of such land, but establishing a processing plant on that land requires agreement from and potential profit sharing with hundreds of others.

The Tainuis' development of their processing plant, a relatively small structure, required considerable financial investments primarily due to the costs of certification. Despite a long history of building their facilities, the Tainuis, seeking the plant's licensure, had to hire certified builders to construct the facility. Compliance with food safety standards required the purchase of extensive plumbing, showers, wash

Figure 1: Location of Te Waihora/Lake Ellesmere on the East Coast of New Zealand's South Island. Map created by Hekia Bodwitch using Canterbury Maps. The imagery and bathymetric data on the map are licensed under Creative Commons 4.0.

stations, and certain forms of gear deemed sanitary. Additional costs were incurred in developing and later operating the plant, including time each day recording and reporting information on various aspects of facility operation. The costs of developing a processing plant, according to the Tainui's plan, were to be offset by selling fish to high-end markets. During the time of plant operation, this possibility was yet to be realized as the Tainuis and others in the Iwi did not have relationships with individuals in these markets. Diverse historical, colonial institutions that restricted Māori access to capital and education (Waitangi Tribunal 1992), necessary to develop relationships with potential buyers overseas, continued to affect Māori fishery development opportunities in ways the Settlement, on its own, did not address.

Access to Fish

The costs of establishing a fishing operation and developing a processing plant might also be offset by the possibility of more fish coming through the plant, either through quota purchases or increased stock numbers. The Tainuis had yet to realize this latter option because non-fishing forms of development have affected their access to fish. Across New Zealand, eel populations have declined due to habitat loss associated with dairy industry expansion. This expansion reflects the lucrativeness of New Zealand's dairy export industry, which compounded following the New Zealand government's signing of a free-trade agreement with China in 2008 (Bodwitch et al. 2022). The impacts of such development on Māori fishers, or the region's waterways more broadly (Joy et al. 2022), are not factored into most accounts of the industry's economic contributions. The Tainuis had to shut down their plant in 2019 because there were no eels. In the Tainui's case and others, upstream agricultural development disproportionately impacts Māori due to the legacies of colonial land removal policies. These policies pushed Māori off lucrative agricultural land to make way for non-Māori farmers. With minimal access to land and lacking capital to access offshore fisheries, Māori relied on downstream, inshore fisheries for subsistence and trade (Waitangi Tribunal 1992).

At Te Waihora, the Ngāi Tahu Iwi holds co-governance rights but does not have the political authority to regulate upstream developments due in large part to the perceived economic contributions of the dairy sector and also due to lack of data on impacts (Bodwitch et al. 2022). The delay between ITQ system establishment and Iwi fishers obtaining fishing rights again posed a problem, in that exponential growth of the dairy industry upstream began when Māori were not fishing commercially. Few local "eyes and ears" were on the ground to collect data that might otherwise inform challenges to upstream development (McCarthy et al. 2013). In this case, lack of knowledge affected the ability to govern. This problem of competition with other forms of resource use, as well as the challenges associated with accessing markets, was not encompassed in the bounds of the problem addressed by the Settlement. Ngāi Tahu also holds rights to regulate fishing pressure on the lake, including the implementation of bans in important fish nursery grounds. Yet, the authority to regulate fishing activity does little to curb the effects of non-fishing threats to resource use. In light of these dynamics, intergenerational Māori eel fisher Colin

Peters has described the government's actions toward Māori fishers as a "war of attrition." Peters did not have the reported catches necessary to obtain quota during the initial allocation process. His testimony informed the 1992 Fisheries Settlement claim, providing evidence of injustices committed against Māori. Yet, as he describes, "Has the Peters family received any benefit from the Fisheries Settlement? No, not one cent."

Discussion and Conclusion: Institutional Polycentricity and Equity

This access analysis has demonstrated how Indigenous rights reconciliation initiatives, those that transfer rights but not also the authority to govern the factors affecting how rights can be used, can be insufficient to advance equity for Indigenous groups (Table 2). Māori obtained commercial fishing rights through the New Zealand government's efforts to reconcile Māori fishery-related grievances. Neither Māori leaders nor fishers obtained the means necessary to develop fisheries in compliance with the state-led institutions governing New Zealand's fisheries. These include, especially, regulations on who can buy and sell fish. Additionally, Māori owners of commercial fishing rights did not obtain autonomy to govern the factors affecting fish population health, including the ability to regulate upstream resource users' activities. The number of Indigenous-held commercial fishing rights in New Zealand is, therefore, disproportionate to the number of Indigenous-owned fishing operations in New Zealand for two key reasons. Firstly, Māori fishers are unable to access lucrative markets. Secondly, Māori fishing rights holders can also not curb upstream drivers of fishery depletion to ensure the ecological stability of a fishery and the economic security of any financial investments in their fishing operations.

This analysis holds implications for environmental governance, Indigenous rights reconciliation initiatives, and fishery development in New Zealand and elsewhere. For scholars of environmental governance, this case illustrates that institutional diversity does not necessarily equal polycentricity, as defined by the ability of citizens to organize into decision-making units (Ostrom et al. 1961). An institutionally diverse governance regime may be monocentric or only partially polycentric. Moreover, Māori fishers' development experiences indicate a need for advocates of institutional diversity to be specific about which aspects of institutionally diverse contexts facilitate desired outcomes. Institutional diversity can inhibit equity when multiple layers of regulations increase costly administrative burdens individuals must overcome to participate in development or governing initiatives, excluding those who are less well-resourced. For Māori fishers, diverse sets of rules governing fish trade increased costs associated with direct fisher-to-consumer marketing, restricting fishers' abilities to sell their own fish. Institutional diversity may also inhibit equity when diversity manifests as distinct and uncoordinated rule-making bodies with inequitable forms of political authority in contexts of shared resource use. In New Zealand, fishers have little authority to influence rules governing upstream farmers, whose actions nonetheless affect fisheries downstream. The governing challenges associated with a lack of coordination between upstream and downstream rulemaking processes are not unique to New Zealand

Table 2: Using an access analysis to examine relationships between institutional diversity, polycentricity, and equity in New Zealand's fisheries.

Phase 1	Phase 2	Phase 3
Identify the conditions that must be met for an individual or group to benefit from a particular resource.	*Identify any dynamics constraining an individual's or group's ability to meet the conditions identified in Phase 1.*	*Identify institutions that mediate the inequitable effects of dynamics identified in Phase 2.*
Access to rights	a. The New Zealand government's initial quota allocation criteria b. Capital to purchase quota	a. 1992 Fisheries Settlement b. Iwi quota allocation initiatives
Access to means of production (i.e., boat, gear, labor)	a. Access to capital (social as well as financial) b. Regulations regarding fishing practices	a. Iwi financial support (i.e., subsidies) for Iwi fishers b. N/A (monocentric governing initiative, no institutions substantially mediate inequitable effects)
Access to markets	a. Relationships with buyers b. Regulations regarding fish processing and trade practices	a. Iwi tribal branding & marketing initiatives (including establishing Iwi processing companies) b. N/A (monocentric governing initiative, no institutions substantially mediate inequitable effects)
Access to fish	a. Other fishers' illicit activities b. Non-fishing-related threats to fish habitat (i.e., upstream land use change)	a. Māori fishery monitoring initiatives in culturally significant fishing areas b. N/A (a co-managed but monocentric governing regime regulates upstream land use; no institutions substantially mediate inequitable effects)

(Song et al. 2018). These challenges, highlighted here, demonstrate the significance of scholars' claims that polycentric regime outcomes are affected by the extent to which learning occurs between governing units (Aligicia 2014).

In order to develop governing systems that advance equity, this case suggests that increasing options for polycentricity, as defined by the ability of citizens to organize to meet their needs, may result in the establishment of more equitable environmental governing regimes. In contexts where Māori have been able to self-govern, Māori fishers' development has been advanced. This has occurred in the allocation of rights, as made possible by the Settlement, and through Iwi initiatives to provide subsidies to Iwi fishers to support their development. For theories of polycentricity, this relationship between the autonomy of Māori groups and fishery development outcomes draws attention to key questions remaining in the field regarding the degree of autonomy necessary to advance alternative outcomes, a concept that has remained poorly defined (Carlisle and Gruby 2019). For example, Carlisle and Gruby (2019) draw on Ostrom et al. (1961) to describe autonomy as "decision-making centers" that "act on their own behalf, without centralized coordination" (Carlisle and Gruby 2019). Others' accounts of the autonomy necessary to advance polycentric systems that achieve particular benefits are similarly vague. Marshall (2009) uses the term "considerable" to describe the types of autonomy actors in polycentric systems hold. Komakech and van der Zaag (2013) describe polycentric systems as comprised of "semiautonomous" actors. Andersson and Ostrom (2008) indicate that actors require "some degree of autonomy" for polycentric systems to function. This lack of specificity provides little help for those attempting to negotiate for rights to advance particular social goals. Access theory (Ribot and Peluso 2003), we suggest, presents a method for adding this specificity by providing a way to unpack the governance regimes that affect how burdens and benefits from natural resources are distributed. Access analyses, in turn, may serve to build an understanding of which groups have (or should have) the autonomy to implement governing strategies to address inequality. As such, and in response to broader debates around fishery governance, the degree of autonomy Indigenous groups must hold to overcome inequitable constraints may involve whatever degree is necessary to ensure historically disadvantaged groups can access the conditions necessary for development or governance.

For analyses of Indigenous rights reconciliation agreements, the findings presented here demonstrate how New Zealand's 1992 Fisheries Settlement did not address barriers to fishery development that Māori fishers have experienced even while holding quota rights, but that this should not be viewed as a limitation of the Settlement itself. Rather, these findings indicate that addressing equity is a dynamic process, as the conditions affecting access change with time, for instance, as markets become consolidated or upstream resource use intensifies. As a result, the processes for reconciling historical or contemporary injustices (in New Zealand and likely elsewhere) are never "full and final," as the New Zealand government claims settlements to be in New Zealand. Instead, because the contexts affecting how assets can be used are always changing, whether assets granted in settlements or other forms of reconciliation agreements are sufficient mechanisms to alter inequitable distributions of burdens and benefits from resources is also always changing. The concept of polycentricity provides a model of governance, as a governance regime in which citizens can organize to address their needs, that allows strategies to respond to this dynamism. Access analyses provide a lens through which to identify it. These findings challenge claims advocated by certain scholars and public commentators, that limited Indigenous development is due to leaders' mismanagement of collectively-owned fishing rights (Rata 2011; De Alessi 2012). This analysis demonstrates how state-implemented regulations around markets and other forms of resource use can continue to exclude Indigenous fishers from accessing benefits from fisheries, even with rights to do so.

In a practical sense, Māori fishers' development experiences indicate strategies that may serve to advance justice for historically disadvantaged groups in the future. In particular, the ways Māori fishers have been supported by Iwi funds raises questions about the equity implications associated with proposals from the World Trade Organization (WTO) to ban fishery subsidies (e.g., Sumaila et al. 2021). In efforts to achieve more equitable distributions of benefits from fishing resources, subsidies that help historically disadvantaged fishers obtain boats or develop processing operations may be necessary. Additionally, the previous suggestion that the degree of autonomy necessary to support equity is the degree that enables groups to overcome barriers to accessing benefits from resources also encourages analysis of non-subsidy-based ways Iwi or other Indigenous groups might support Indigenous fishers' development. Such efforts may include strategies that reduce the administrative burdens fishers must overcome to access markets, as might occur with Indigenous-governed fish trade initiatives, that allow fishers to sell directly to consumers without adhering to expensive government regulations for licensing fish sellers. Direct fisher-to-consumer initiatives have been advanced in various contexts (Gillette and Vesterberg 2022), with licensing and reporting requirements identified as a key limitation (Stoll et al. 2022). An ongoing analysis of how these initiatives help fishers access benefits from resources will aid in understanding their suitability for elsewhere. In New Zealand, fish trade regulations, a key barrier to Māori fishery development, were designed to ensure compliance with the nation's ITQ system as well as food safety standards. However, it is possible that Indigenous or other groups of fishers have strong enough social bonds to self-monitor compliance without needing additional government oversight. Further, regarding policy that supports equitable fishery development at the land-sea interface, establishing multiple autonomous, Indigenous, and state authorities for governing transboundary resources may advance the changes necessary to curb the inequitable effects of upstream farming practices on downstream fishers. This is especially likely to be the case if non-Indigenous leaders are exposed to different political pressures to support non-Indigenous farmers than Indigenous leaders face (i.e., Bodwitch et al. 2022).

In conclusion, the challenges associated with institutional diversity as a potential barrier to equity are not unique to New Zealand. The ways administrative burdens, resulting from multiple layers of regulations, can inhibit the development of historically disadvantaged groups have also been documented as barriers to small-scale producers' abilities to survive in other development contexts, including agriculture (Bodwitch et al. 2021). Although institutional diversity can mitigate risks associated with failed governing initiatives, diverse sets of rules and policies can also present barriers to participation when overseen by monocentric systems, in which disadvantaged citizens have minimal authority to innovate alternatives. In these contexts, new models of governance that grant underserved groups degrees of autonomy necessary to overcome inequitable regulatory hurdles are warranted. Polycentricity, then, in this conception, manifests as a system through which citizens can experiment with initiatives that reduce costly administrative burdens necessary to participate in certain forms of development, in addition to experimenting with context and culturally relevant rules for resource use. Access analyses can help evaluate whether desired benefits are achieved as these initiatives unfold.

Acknowledgments

The authors thank the interview participants, especially Te Waihora fishers, who generously donated their valuable time, experience, and knowledge to the study. We also thank the constructive feedback of the editors and three anonymous reviewers. Additionally, we gratefully acknowledge support from a Canada First Research

Excellence Fund grant through the Nippon Foundation Ocean Frontier Institute and the Ocean Nexus Center at the University of Washington EarthLab. We also acknowledge support from the New Zealand Ministry of Business, Innovation, and Employment's Unlocking Export Prosperity Programme [LINX1701].

References

Acton, L., R.L. Gruby and 'A. Nakachi. 2021. Does polycentricity fit? Linking social fit with polycentric governance in a large-scale marine protected area. J. Enviro. Manage. 290(112613). https://doi.org/10.1016/j.jenvman.2021.112613.

Aligica, P.D. 2014. Institutional Diversity and Political Economy: The Ostroms and Beyond. Oxford University Press, New York.

Anaya, J. 2011. Report of the Special Rapporteur on the rights of Indigenous peoples: The situation of Māori people in New Zealand. United Nations Human Rights Council. A/HRC/18/35/Add.4.

Andersson, K.P. and E. Ostrom. 2008. Analyzing decentralized resource regimes from a polycentric perspective. Policy Sci. 41(1): 71–93. https://doi.org/10.1007/sl.

Boast, R.P. 1999. Māori Fisheries 1986-1998: A Reflection. VUWLR. 30(1): 111–134.

Bodwitch, H. 2014. Why feminism? How feminist methodologies can aid our efforts to "give back" through research. J. Res. Pract. 10(2): 1–9.

Bodwitch, H. 2017a. Challenges for New Zealand's individual transferable quota system: Processor consolidation, fisher exclusion, & Māori quota rights. Mar. Policy. 80: 88–95. https://doi.org/10.1016/j.marpol.2016.11.030.

Bodwitch, H. 2017b. Property is not sovereignty: Barriers to indigenous economic development in Aotearoa/New Zealand's fisheries. UC Berkeley. ProQuest ID: Bodwitch_berkeley_0028E_17400. Merritt ID: ark:/13030/m5tb63tj. Retrieved from https://escholarship.org/uc/item/7hq099dr.

Bodwitch, H., M. Polson, E. Biber, G.M. Hickey and V. Butsic. 2021. Why comply? Farmer motivations and barriers in cannabis agriculture. J. Rural Stud. 86: 155–170. https://doi.org/10.1016/j.jrurstud.2021.05.006.

Bodwitch, H., A.M. Song, O. Temby, J. Reid, M. Bailey and G.M. Hickey. 2022. Why New Zealand's Indigenous reconciliation process has failed to empower Māori fishers: Distributional, procedural, and recognition-based injustices. World Dev. 157(105894). https://doi.org/10.1016/j.worlddev.2022.105894.

Bodwitch, H., K.M. Hamelin, K. Paul, J. Reid and M. Bailey. 2024. Indigenous self-determination in fisheries governance: implications from New Zealand and Atlantic Canada. Front. Mar. Sci. 11: 1297975. http://doi.org/10.3389/fmars.2024.1297975.

Campbell, D. 1984. Individual transferable catch quotas: their role, use and application. Australian Northern Territory, Department of Primary Production, Fishery Report 11, Darwin, Australia.

Carlisle, K. and R.L. Gruby. 2019. Polycentric systems of governance: a theoretical model for the commons. Policy Stud. J. 47(4). https://doi.org/10.1111/psj.12212.

Carlisle, K.M. and R.L. Gruby. 2018. Why the path to polycentricity matters: evidence from fisheries governance in Palau. Env. Pol. Gov. 28: 223–235. https://doi.org/10.1002/eet.1811.

Carothers, C. and C. Chambers. 2012. Fisheries privatization and the remaking of fishery systems. Environ. Soc. Adv. Res. 3: 39–59. https://doi.org/10.3167/ares.2012.030104.

Christy, F.T. 1973. Fisherman quotas: a tentative suggestion for domestic management. Occasional Paper 19. Kingston: Law of Sea Institute, University of Rhode Island.

Copes, P. 1986. A critical review of the individual quota as a device in fisheries management. Land Econ. 62(3): 278–291.

De Alessi, M. 2012. The political economy of fishing rights and claims: the Māori experience in New Zealand. J. Agrar. Chang. 12: 390–412.

De Wit, F. and J. Mourato. 2022. Governing the diverse forest: Polycentric climate governance in the Amazon. World Dev. 157: 105955. https://doi.org/10.1016/j.worlddev.2022.105955.

Diver, S., M.V. Eitzel, S. Fricke and L. Hillman. 2022. Networked Sovereignty: Polycentric Water Governance and Indigenous Self-determination in the Klamath Basin. Water Altern. 15(2): 523–550.

Engle, K. 2010. The Elusive Promise of Indigenous Development: Rights, Culture, Strategy. Duke University Press, Durham and London.

Gillette, M.B. and V. Vesterberg. 2022. Dead in the water? Sustainability and direct seafood sales in Sweden. J. Rural Stud. 89: 248–256.

Haraway, D. 1988. Situated Knowledges: The Science Question in Feminism and the Privilege of Partial Perspective. Fem. Stud. 14(3): 575–599.

Harding, S.G. 2004. The feminist standpoint theory reader: Intellectual and political controversies. Routledge, New York.

Holifield, R., J. Chakraborty and G. Walker. 2018. The worlds of environmental justice. pp. 1–11. *In:* Holifield, R., J. Chakraborty and G. Walker (eds.). The Routledge Book of Political Ecology, Routledge.

Joy, M.K., D.A. Rankin, L. Wöhler, P. Boyce, A. Canning, K.J. Foote et al. 2022. The grey water footprint of milk due to nitrate leaching from dairy farms in Canterbury, New Zealand. Australas. J. Environ. Manag. 29(2): 177–199. https://doi.org/10.1080/14486563.2022.2068685.

Komakech, H.C. and P. van der Zaag. 2013. Polycentrism and Pitfalls: The Formation of Water Users Forums in the Kikuletwa Catchment, Tanzania. Water Int. 38(3): 231–49.

Marshall, G. 2009. Polycentricity, Reciprocity, and Farmer Adoption of Conservation Practices under Community-Based Governance. Ecol. Econ. 68: 1507–20.

McCarthy, A., C. Hepburn, N. Scott, K. Schweikert, R. Turner and H. Moller. 2013. Local people see and care most? Severe depletion of inshore fisheries and its consequences for Māori communities in New Zealand. Aquat. Conserv. https://doi.org/10.1002/aqc.2378.

McCormack, F. 2010. Fish is my daily bread: owning and transacting in Māori fisheries. Anthropol. Forum. 20(1): 19–39. https://doi.org/10.1080/00664670903524194.

McCormack, F. 2021. Interdependent Kin in Māori Marine Environments. Oceania. 91(2): 197–215. https://doi.org/10.1002/ocea.5308.

Memon, P.A.U. and R. Cullen. 1992. Fishery Policies and their Impact on the New Zealand Māori. Mar. Resources Econ. 7: 153–167.

Mohai, P., D. Pellow and J.T. Roberts. 2009. Environmental justice. Annu. Rev. Environ. Resour. 34(405–30): 405–430. https://doi.org/10.1146/annurev-environ-082508-094348.

Moynihan, D., P. Herd and H. Harvey. 2014. Administrative burden: learning, psychological, and compliance costs in citizen-state interactions. J. Publ. Adm. Res. Theor. 25: 43–69. https://doi.org/10.1093/jopart/m.

Morrison, T.H. 2017. Evolving polycentric governance of the Great Barrier Reef. PNAS 114: E3013–E3021. https://doi.org/10.1073/pnas.1620830114.

Myers, R. and C.P. Hansen. 2020. Revisiting A Theory of Access: A review. Soc. Nat. Resour. 33(2): 146–166. https://doi.org/10.1080/08941920.2018.1560522.

Orange, C. 2004. An Illustrated History of the Treaty of Waitangi. Bridget Williams Books Ltd, Wellington, New Zealand.

Ostrom, E. 2009. A Polycentric Approach for Coping with Climate Change. World Bank Policy Research Working Paper No. 5095.

Ostrom, E. 2010. Polycentric systems for coping with collective action and global environmental change. Glob. Environ. Change. 20(4): 550–557. https://doi.org/10.1016/j.gloenvcha.2010.07.004.

Ostrom, E. 2012. Why do we need to protect institutional diversity?*. Eur. Political Sci. 11: 128–147. https://doi.org/10.1057/eps.2011.37.

Ostrom, V., C.M. Tiebout and R. Warren. 1961. The organization of government in metropolitan areas: A theoretical inquiry. Amer. Polit. Sci. Rev. 55(4): 831–842. https://doi.org/10.1017/S0003055400125973.

Pattberg, P. and O. Widerberg. 2016. Transnational multistakeholder partnerships for sustainable development: Conditions for success. Ambio. 45: 42–51. https://doi.org/10.1007/s13280-015-0684-2.

Parsons, M., L. Taylor and R. Crease. 2021. Indigenous environmental justice within marine ecosystems: a systematic review of the literature on indigenous peoples' involvement in marine governance and management. Sustainability. 13(4217): 1–33. https://doi.org/10.3390/su13084217.

Peluso, N.L. and J. Ribot. 2020. Postscript: A Theory of Access Revisited. Soc. Nat. Resour. 33(2): 300–306. https://doi.org/10.1080/08941920.2019.1709929.

Pinkerton, E. and R. Davis. 2015. Neoliberalism and the politics of enclosure in North American small-scale fisheries. Mar. Policy. 61: 1–10. https://doi.org/10.1016/j.marpol.2015.03.025.

Rahman, H.M., A. Saint Ville, A. Song, J. Po, E. Berthet, J. Brammer et al. 2017. A framework for analyzing institutional gaps in natural resource governance. Int. J. Commons. 11(2): 823–853. https://doi.org/10.18352/ijc.758.

Rata, E. 2011. Encircling the commons: Neotribal capitalism in New Zealand since 2000. Anthropol. Theory. 11(3): 327–353. https://doi.org/10.1177/1463499611416724.

Ribot, J.C. and N.L. Peluso. 2003. A theory of access. Rural Sociol. 68(2): 153–181. https://doi.org/10.1111/j.1549-0831.2003.tb00133.x.

Rout, M., B. Lythberg, J. Mika, A. Gillies, H. Bodwitch, D. Hikuroa et al. 2019. Kaitiaki-centred business models: case studies of Māori marine-based enterprises in Aotearoa New Zealand. Wellington, New Zealand: Sustainable Seas National Science Challenge.

Rout, M., J. Reid and J. Mika. 2020. Māori agribusinesses: the whakapapa network for success. AlterNative. 16(3): 193–201. https://doi.org/10.1177/1177180120947822.

Sissenwine, M.P. and P.M. Mace. 1992. ITQs in New Zealand: the era of fixed quota in perpetuity. Fish. Bull. 90: 147–160.

Song, A.M., S.D. Bower, P. Onyango, S.J. Cooke, S.L. Akintola, J. Baer et al. 2018. Intersectorality in the governance of inland fisheries. Ecol. Soc. 23(2): 17. https://doi.org/10.5751/ES-10076-230217.

Steffen, W., J. Rockström, K. Richardson, T.M. Lenton, C. Folke and D. Liverman. 2018.Trajectories of the Earth System in the Anthropocene. PNAS. 115(33): 8252–8259. https://doi.org/10.1073/pnas.1810141115.

Stoll, J.S., H.K. Harrison, E. De Sousa, D. Callaway, M. Collier, K. Harrell et al. 2021. Alternative Seafood Networks During COVID-19: Implications for Resilience and Sustainability. Front. Sustain. Food Sys. 5(614368): 1–12. https://doi.org/10.3389/fsufs.2021.614368.

Sumaila, U.R. 2010. A Cautionary Note on Individual Transferable Quotas. Ecol. Soc. 15(3).

Sumaila, U.R., D. Skerritt, A. Schuhbauer, S. Villasante, A.M. Cisneros-Montemayor and H. Sinan. 2021. WTO must ban harmful fisheries subsidies. Science 374(6567): 544. https://doi.org/10.1126/science.abm1680.REF.

Urlich, S.C., F.R. White and H.G. Rennie. 2022. Characterizing the regulatory seascape in Aotearoa New Zealand: Bridging local, regional and national scales for marine ecosystem-based management. Ocean Coast. Manag. 224(106193). https://doi.org/10.1016/j.ocecoaman.2022.106193.

Waitangi Tribunal. 1992. Ngai Tahu Sea Fisheries Report. New Zealand Government Department of Justice. Wai 27. Waitangi Tribunal. 2016. Report on Claims about the Reform of Te Ture Whenua Māori Act 1993. New Zealand Government Department of Justice. Wai 2478.

Yandle, T. and C.M. Dewees. 2008. Consolidation in an individual transferable quota regime: lessons from New Zealand, 1986-1999. J. Environ. Manage. 41(6): 915–928. https://doi.org/10.1007/s00267-008-9081-y.

Young, O.R., D.G. Webster, M.E. Cox, J. Raakjær, L.Ø. Blaxekjær, N. Einarsson et al. 2018. Moving beyond panaceas in fisheries governance. PNAS. 115: 9065–9073. https://doi.org/10.1073/pnas.1716545115.

6

Institutional Change Between the RastafarI Movement and the Formal State in Jamaica
A Historical Perspective

Arlette Saint Ville,[1,*] *June Y. T. Po*[2] *and Amilcar Sanatan*[3]

Introduction: Reconnecting Discourses of Agriculture, Food, and Land with Local Institutions and People

Numerous international donor and state-driven agriculture, nutrition, and food security programs and policies have been initiated in response to the decline in the domestic agri-food system in the Caribbean. However, there are significant concerns about persistent malnutrition and hidden fragilities in the agri-food sector exacerbated by shocks and stresses (Hickey and Unwin 2020; Lowitt et al. 2016). Previous research on the relationship between local smallholder farmers, consumers, communities, and international or state-driven programs in the context of small island developing states (SIDS) notes that development programs that pay insufficient consideration to the value of context-specific social structures were unlikely to have lasting positive impacts (Baviskar 2003). Social structures, history, and values vary across contexts and give rise to differences in rule formulation and social norms among groups. These rules and norms comprise institutions defined by North (1991, 97) as "humanly devised constraints that structure political, economic and social interactions" and

[1] Department of Geography, The University of the West Indies, St. Augustine Campus, Trinidad and Tobago.

[2] Natural Resources Institute, University of Greenwich, Medway Campus, Central Avenue, Chatham Maritime, Kent ME4 4TB, UK.
 Email: J.Y.T.Po@gre.ac.uk

[3] Department of Literary, Cultural and Communication Studies, The University of the West Indies, St. Augustine Campus, Trinidad and Tobago.
 Email: amilcar.sanatan@my.uwi.edu

* Corresponding author: arlette.saintville@sta.uwi.edu

have become a key arena for attention to understand institutional diversity better. As such, understanding institutions can reveal how their development and interactions shape social-ecological systems to promote or undermine sustainability (Berkes and Folke 1998). This suggests that a food-secure and sustainable Caribbean food future will need to understand better institutional diversity to support a healthier society with more decentralized, socially and ecologically appropriate, adaptive, and heterogeneous institutions inclusive of grassroots discourses and actions.

Despite the need for context-specific grassroots discourses in the agri-food system, much of the food-related practice and research are informed by institutions crafted from a top-down, productivist paradigm, in part due to colonial legacies coming out of the export-focused agricultural history in the Caribbean (Saint Ville et al. 2015). When a productivist paradigm sets the tone and narratives in the agri-food system, it primarily promotes food trade, technological advancement, and supply-side interventions (Béné et al. 2019), with projects often funded by donor agencies. While such projects provide resources to underfunded sectors in low- and middle-income countries (LMIC), they may be out of sync with community-level institutions. Further, modern theories that influence state-building are informed by utilitarian and neo-classical economic perspectives for decision-making with little emphasis on context-specific social and cultural practices of SIDS. It is of concern that bottom-up processes that link communities and place-based narratives with the ongoing development of communities of practice (Lowitt et al. 2015) have had relatively little attention and support in the SIDS context. We argue that engagement with communities and understanding local governance is critical to reconnecting discourses of agriculture, nutritious food, and land with intersecting local issues such as people's livelihoods and dignity (Rodríguez et al. 2017).

There has been increased interest in unpacking how food system governance contributes to social and environmental sustainability. Not only do the architecture of rules and social norms within local and global food systems serve to guide social actors in managing social and natural resources in their world (Ostrom et al. 1994), but maladaptive governance based on outdated, ill-devised or imposed social structures and norms can have detrimental consequences on natural resource stocks and ecosystem services. Rule development may be informed by varied place-specific factors such as socioeconomic conditions, cultural practices, geographical settings, characteristics of the resource system, state of knowledge, market mechanisms, requirements of globalization, and climate change adaptations. Jayaprakash and Hickey (2019) highlight the importance of appreciating how historical factors weave into contemporary institutional change. They explain how legacies of past social structures (caste in their case) remain a visible presence in the landscape and affect institutional functioning in post-colonial India. Further, the authors claim that the resulting landscape can be understood as a mosaic built up from interactions between legacies, actors, and nature. This suggests that while rules provide signposts that direct actors and outcomes (Ostrom 1999), understanding and mapping these changing rules and interactions over time play a critical role in advancing or limiting social and environmental sustainability within social-ecological systems where diverse institutions are at play (Ostrom 2005; Po et al. 2019).

Sustainability is understood to comprise interdependent environmental, economic, and social elements. Of these three, the concept of social sustainability related to food is the most underdeveloped (Toussaint et al. 2021). It includes access to critical social and other resources, human and labor rights, dignity, food security, and meaningful livelihoods for all actors across the value chain while sustaining a thriving natural resource base needed for the present life and for future generations. Along with understanding this key social dimension of sustainability, there is a need to add to the existing body of evidence with a historical lens and analysis to inform place-based discourse and practice for sustainability in natural resource governance. This chapter explores the institutional interactions and agri-food system developments in the Caribbean. Drawing on the literature from geographies of food (Whatmore and Thorne 1997; Winter 2005) and the interrelated concepts of the right to food and cultural rights of people (International Covenant on Economic, Social and Cultural Rights 1966), we examine the socio-political and historical process of the RastafarI movement. The RastafarI movement's growth and interactions in relation to the formal state of Jamaica in the English-speaking Caribbean are framed against the backdrop of the global food sovereignty movement. Jamaica is a typical case of a post-colonial English-speaking society grappling with its nascent national reconstructions of the agri-food system.

In this chapter, we first trace the historical context of the English-speaking Caribbean agri-food system. Second, we introduce the development of the global food sovereignty discourse. Third, we introduce the socio-political history of the RastafarI movement in Jamaica and their philosophical, ecological, and social approach to natural living and food. Fourth, we highlight the stewardship ethic that informs their human-nature relationship. Fifth, we offer a historical analysis to trace their interactions over time with the formal state of Jamaica to highlight key developments. Finally, we center the discussion on their institutional development in the struggle for racial and economic distributive and rights-based justice in relation to food over the past decades. Through the struggle of the RastafarI movement as a historical case study of locally-based institutional diversity and change, this chapter aims to demonstrate its inextricable connections to contested issues of human and land rights and racial, social, ecological, economic, and food justice.

Post-Colonial Caribbean Food Sovereignty Discourse: From Colonial Legacies to "Livity"

"Caribbean food, cuisine, official languages, and Creole dialects are also functions of these colonial encounters" (Brissett 2018).

Colonial Legacies of the English-Speaking Caribbean's Agri-Food System

While history books describe how, from the mid-1500s to the mid-1700s, islands of the Caribbean were covered with sugar cane plantations and mill refineries, they often leave readers unappreciative of "versions of systems of colonizer nations" that remain (Brissett 2018). Historical accounts describe the agroecological characteristics of fertile soil, relatively flat land near coastal waters, the trans-Atlantic slave trade that forcefully displaced West African peoples, and the cultivation of sugar cane as the most important crop throughout the Caribbean (Watts 1990). Jamaica was one of the world's main sugar producers in the mid-1700s (Warsh 2014). Sugar (agri)-culture has deeply impacted society and the economy with significant environmental consequences. Deforestation, water pollution, and soil degradation at differing scales across the region occurred during the colonial and post-colonial eras (Richardson 1992). Political independence from Europe did not end the unsustainable social and ecological values that were based on racial and economic inequities. Instead, these values remained embedded in "the fabric of Caribbean social and ecological identities" (Brissett 2018).

Caribbean colonial society was heavily agrarian, yet malnutrition was widespread. This contradiction "was not a reflection of any inherent human or bio-physical constraints to food self-sufficiency but rather was an outcome of the distortions of plantation economies (Beckford 1972)" (Weis 2003). Colonizers framed the Caribbean as the "sugar bowl, tobacco pouch, coffee shop and rum supplier of the world." In order to serve their imperial interests (Mintz 1985, 130), colonial plantation economies misdirected local productive capacities away from local needs. Dualities of plantation and peasant agriculture remained as dominant features of Caribbean economies up until the 20th century (Saint Ville et al. 2015), when other economic sectors, such as tourism, emerged. Timms (2008, 102) observes, "[p]roduction of foodstuffs in the colonies was discouraged as resources were focused on tropical export products while sustenance needs were met through the importation of agricultural products produced in the temperate regions of the metropole."

While consensus on these economically important issues may have existed in England, they were politically charged in the colonies. In the 1950s and 60s, these issues and their framings were sites of heated cultural exchanges in the soon-to-be independent states aiming to devise new national policies. Around this time in the newly independent Caribbean, the Plantation Economy Model was a school of thought developed by Caribbean intellectuals that fits into the broader school of Dependency Thought (Girvan 2005). Generally, its proponents argue that export-propelled economies served the interests of colonizer economies and established economic dependency in the Caribbean (Pantin 1980). Across the region, since the 1960s, governments have pursued policies that attracted foreign capital for industrialization.

Colonial legacies of the plantation economy sustained export agriculture, monoculture production, and food imports. Saint Ville et al. (2015) describe how these path-dependent policies were maintained by rigid plantation-era institutions with less attention to the domestic production sector. While formally committing to economic diversification, it has been noted that "the neglect of domestic agriculture has resulted in the Caribbean becoming a net food importer with a growing food import bill and increased food insecurity" (Timms 2008). Despite the theoretical plantation model, which supported breaking away from food trade dependencies with Euro-American economies (Beckford 1972), efforts to re-frame historical relations of power around food and agriculture have not been achieved.

Movements for Food Sovereignty in Latin America and Beyond

By the late 1940s, peasant groups and communities with similar colonial food histories around the world began to demand control over the way food was produced, traded, and consumed. In 1947, the UN Commission on Human Rights (Fairbairn 2010) contributed to the post-war food regime. Over four decades later, the concept evolved to rights-based arguments such as "right to food" and "right to freedom from hunger" (Eide 1996; Wittman 2010). In 1993, grassroots peasants' movements coalesced, forming La Vía Campesina (LVC), "an international movement bringing together millions of peasants, small and medium size farmers, landless people, rural women and youth, indigenous people, migrants and agricultural workers from around the world. This global movement defends peasant agriculture and promotes food sovereignty as a way to ensure social justice and dignity and strongly opposes corporate-driven agriculture that destroys social relations and nature" (LVC 2020). With a representation of between 200 to 500 million members through 167 organizations worldwide (LVC 2019; Dekeyser et al. 2018), La Vía Campesina is one of the most prominent actors and driving forces in the global food sovereignty movement. The movement is evolving and has grown to include the agenda of the landless, urban food insecure, and consumers (Chaifetz and Jagger 2014).

Through a series of international conferences, starting with the Tlaxcala Conference in 1996, delegates from various sectors and social movements worldwide met to discuss and agree upon a common understanding of food sovereignty that was captured in the Nyéléni Declaration on Food Sovereignty in 2007. Embedded within the declaration are the six pillars of food sovereignty most widely used in the literature: (1) food for people; (2) value of food providers; (3) localized food systems; (4) need for local control (territory, land, water, seeds, etc.); (5) building knowledge and skills; and (6) working with nature (see Nyéléni Declaration 2007 for details).

While the Nyéléni Declaration places emphasis on "new social relations" around gender equality, it insufficiently addresses the realities of the typical patriarchal patterns of farming families (Patel 2009) and their production capacity (Haugen 2009; Patel 2009; Beuchelt and Virchow 2012). Moreover, the call to localize food production, trade, and consumption would limit the range of commodity options consumers have come to expect. Diversity in geographical locations and agronomic conditions, food costs, and economic viability in relation to comparative production advantages have been used by critics to counter the underpinnings of food sovereignty (Godfray et al. 2010; Meadows et al. 2009; Cline 2007). Similarly, localized food production and trade arrangements that are often championed by environmental and food movements could undermine the current global food trade. This is an important point since global trade is needed in more than 50 low-income, food-deficit countries (FAO 2010). These challenges and limitations point to the need for continued re-imagining and fine-tuning of the food sovereignty concept, informed by local contexts, to guide public policymakers at local, regional, and global levels (Haugen 2009). An institutional lens may provide a better understanding of food systems governance, rule development, and monitoring and enforcement at local and global levels.

Rastafarl Movement, Social and Ecological Sustainability, and Food Sovereignty in the Caribbean

Food sovereignty embedded in the social and environmental sustainability of the English-speaking Caribbean has a long history. In the pre-Columbian era, Indigenous People in small communities across the tropical and subtropical zones (Nair 1993) were guided by customary institutions, which included a stewardship ethic, self-provisioning, and communal norms. Production took place in "food forests" based on agroforestry principles designed to "mimic a forest in terms of the vertical configuration of plants of different heights" and featured 40 to 50 diverse plant species (Beckford and Campbell 2013). Colonizers extirpated Indigenous Peoples, replaced them with enslaved Africans owned by European masters, and transformed both food and primary forests into sugar plantations for export. New institutions facilitated this transformation into the Caribbean plantation economy, which included commodified land, centralized authoritarian control, substantial sunk financial investment in equipment and technology, forced or coerced labor, foreign market focus, and foreign capital support (Saint Ville et al. 2015; Richardson 1992).

In August 1833, with the passing of the Slave Emancipation Act, all enslaved Africans in the English-speaking Caribbean were granted their freedom. Emancipation caused an exodus of the formerly enslaved from plantations to set up new communities and smallholder production livelihoods. However, for many smallholder farmers, norms and rules during the plantation era remained, dominating land use, politics, economy, and society (Saint Ville et al. 2015). National independence saw little structural change in society's dualistic characteristics. In the case of Jamaica, large-scale export production informed by "economic growth dogma" privileged monoculture crops (such as bananas, citrus, coffee, and sugar), while small-scale production rules guided more traditional small-scale farming of polyculture crops (such as yams, breadfruit, pumpkin, dasheen, and tannia) that operated in parallel (Beckford et al. 2007).

Out of and in response to the oppressive, externally-oriented, colonial social order of Jamaican society in the 1930s, Rastafarl emerged (Edmonds 2003). The group was comprised mainly of working-class Jamaicans of African descent. They critiqued the racist and class-based organization of society, the denigration of people of African descent, and the limits of nationalist politics that advanced projects based on "Western modernity" after Caribbean independence (Forsythe 1980). Further, working together in communes oriented around self-provisioning, Rastafarl developed new rules and social norms. As a way of life, a philosophy, and an ethical code, "the Rastafarl movement sought through primarily non-violent struggle... to regain the sense of personal worth and dignity" that was denied by the colonial society (Sampson 1985).

Emerged as a rights-based collectivity, Rastafarl's values and practices toward environmental stewardship existed well before the more recently developed Nyéléni Declaration on food sovereignty. As shown in Table 1, Rastafarl philosophy and practices align with global food sovereignty principles found in the Nyéléni Declaration (2007). A number of key ideas in contemporary Rastafarl have influenced their views and practices on holistic living and ecological sustainability. In particular, "Livity," a central Rastafarl spiritual concept, encompasses several principles

Table 1: Alignment of RastafarI philosophy and practices with global food sovereignty principles found in the Nyéléni Declaration.

6 Pillars Of Food Sovereignty (Nyéléni Declaration 2007)	Examples of Rastafari Environmental Stewardship Ethic
1. Focuses on Food for People: Sufficient, healthy, and culturally appropriate food for all individuals, peoples, and communities	RastafarI **promote and consume** *Ital,* **which is best understood as "pure and whole"** foods, free of preservatives. RastafarI practice food self-provisioning (Sampson 1985). The Ital diet is informed by the Levitical laws from the Old Testament and aligns with the NOVA classification on unprocessed foods (Monteiro et al. 2018).
2. Values Food Providers: Values and supports the contributions and respects the rights of women and men, peasants and small-scale family farmers	Originally working poor, RastafarI **value people over profit** and seek to break the spirit of competition and individualism, which permeated society and its main institutions (Campbell 1987; Barcant 2014).
3. Localizes Food Systems: Brings food providers and consumers closer together; puts providers and consumers at the center of decision-making	RastafarIans maintain food self-sufficiency and learn to work on the land in part to minimize dependencies on imported food purchases. RastafarIans are the food producers, providers, and consumers, creating an ultra-localized food system when possible.
4. Puts Control Locally: Places control over territory, land, grazing, water, seeds, livestock, and fish populations on local food providers and respects their rights	The movement honors individual rights within their collective organizing. In their practices, **localization of control and respect for individual rights and dignity** is a fundamental departure from bureaucracies or hierarchies of the formal state (Edmonds 2003).
5. Builds Knowledge and Skills: Builds on the skills and local knowledge of food providers and their local organizations that conserve, develop, and manage localized food production	RastafarI understand the power of knowledge and the need for ongoing social and spiritual dialogue through 'reasoning' (Campbell 1987). An Afro-centric **consciously molded language 'Iyaric'** was developed (Barcant 2014; Cashmore 1983).
6. Works With Nature: Uses the contributions of nature in diverse, low external input agroecological production and harvesting methods that maximize the contribution of ecosystems and improve resilience and adaptation, especially in the face of climate change	The essence of being RastafarI and practicing Livity includes low-input farming, living close to and as a **part of nature, and** holistic ecology.

Source: Nyéléni Declaration 2007

around a practical lifestyle or "lifeways" centered on naturalness and natural living (Bamikole 2017; Homiak 2005; Gjerset 1994). While Livity has broader personal and social dimensions, it acknowledges the supremacy of all life and the need to preserve, respect, and protect human life, nature, animals, and the natural environment. RastafarI's beliefs emphasize an integrated approach to wellness, focusing on holistic ecology that involves acquiring their own land and land rights to produce their food. They value the consumption of pure food, a diet called "Ital." Ital is derived from RastafarI language ("Iyaric") comprising a combination of the words "vital" and the affirmative "I". The 'Ital' diet involves a mode of organizing food that is organic, vegan, and collectively produced with the dignity of the producer, in contrast to the imported "cheap food" that has resulted from a wage-dependent export economy and colonial legacies.

Although RastafarI philosophies and contributions to the anti-colonial discourses on the decolonization of development have been recognized in academic studies, RastafarI's contribution to the geographies of food in the Caribbean and context-based understanding that intersects planetary health, justice, and human wellbeing have not been well discussed. This intersection of social and environmental sustainability with human rights, food sovereignty, and community building by the RastafarI meets the call for local-level reorganizing of food systems to accomplish ecological sustainability and meet the food needs of local populations (Smith et al. 2017). In the context of SIDS, it has been suggested that efforts toward social and ecological change must be predicated on righting historical wrongs, reorganizing (regionally and globally), and fostering partnerships (Brisset 2018). Through local-level reorganizing, RastafarI has integrated historical issues of racial and social justice, interwoven with Livity, food sovereignty, and ecological sustainability.

Stages in the Institutional Interactions Between RastafarI and the Formal State

From this historical analysis, we illustrate the nature and changing interactions between the RastafarI and the formal state (blue boxes) against the backdrop of the global food sovereignty discourse (green boxes). In the timeline (Figure 1), we traced four stages in the RastafarI-formal state interaction: (1) Growth, (2) Resistance, (3) Co-opted, and (4) Reconciliation.

Growth

Drawing on Marcus Garvey's teachings, the movement was initially organized around the basic religious beliefs centered around racial empowerment and repatriation to Africa among working-class Jamaicans of African descent. Sampson (1985) described this early phase as highly racialized and centered around seven main beliefs: (1) moral fall that led to forced exile (as enslaved people) to the Caribbean; (2) wickedness of Whites (enslavers) as morally inferior to Blacks; (3) impossibility of survival in the Jamaica state; (4) Ethiopia as the promised utopia; (5) Emperor Haile Selassie of Ethiopia as God; (6) eminent repatriation to Ethiopia; and (7) future retribution on Whites. Initially propelled by charismatic leaders and militants, their rhetoric appealed to the poor and disenfranchised urban Black youth

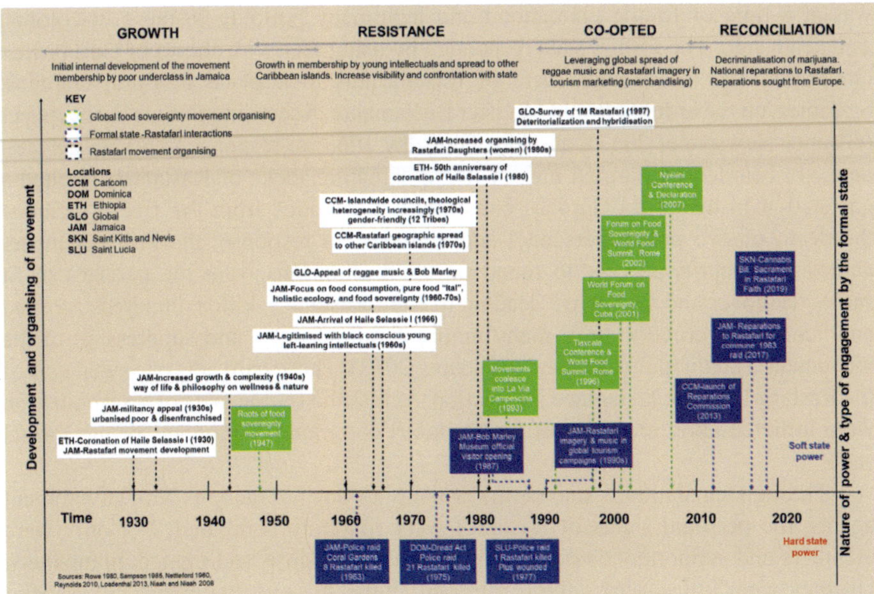

Figure 1: Timelines and stages in the institutional interactions between RastafarI and the formal state.

living in Jamaica's ghettos (Campbell 1987). Over time, other issues emerged, and as a result, RastafarI's appeal and inclusiveness broadened. By the 1960s, Black-conscious, young, left-leaning intellectuals, called "Functional RastafarIans" entered the movement. Their entry, with their higher educational background and socioeconomic status and networks, served to legitimize the movement in Jamaica. By the 1960s and 70s, the growing global appeal of reggae music by noted RastafarI, Bob Marley, led to a "RastafarI Aesthetic" that facilitated the geographic spread of their ideals to other Caribbean islands and countries. With this spread, a decentralized governance model developed in the 1970s with island-wide councils in groups such as the "Twelve Tribes," which increased their theological heterogeneity (Barnett 2005). These developments sustained the movement, and it became globalized over the past 80 years as a spiritual, socio-cultural, and political movement.

Resistance

Disputing societal rules and social norms that revered intersecting race, class, and political dominations, RastafarI came together in decentralized communes designed around new rules that rejected colonial hegemony. The formal state initially charged members as dangerous, in part linked to their norms of anti-colonial language development, marijuana smoking, dreadlocked hair grooming, and off-grid communal living. These norms also served as basic rules of entry to the movement. Members' limited economic resources and disconnection from mainstream society allowed for easy infiltration by working-class criminals needing a hideout (McKeon 2017). However, the charges of treason by the colonial state against the movement are instructive. Philosophically and practically, the state recognized this spiritual, socio-cultural, and political movement as subversive in its intent. This challenge was at a time of fragile nationhood and legitimacy building in the post-colonial Jamaican state. The formal state responded by using fear with shows of hard power. On March 16, 1963, a massacre of RastafarIans took place in Coral Gardens, St. James, on the order of Prime Minister Bustamante. According to a local newspaper (Gleaner Newspaper 2017), on Holy Thursday 1963, in response to a botched land deal, six bearded and armed men (without signature locks of RastafarI) torched a gas station in the Coral Gardens community. Casualties from the fracas included the death of two police personnel and civilians. In response, the Prime Minister authorized security forces "to round up all RastafarIans from the parishes of St James, Hanover and Trelawny," leading to eight RastafarIs killed, hundreds arrested and "collective crucifixion of many innocent followers ... and suppression of the movement islandwide" (Gleaner Newspaper 2017). These hard acts of power against the RastafarI did not take place in isolation. Across the region, acts of discrimination were inflicted upon members of the RastafarI community by agents of the formal state.

Through legislation, social norms, and day-to-day interactions with government agents, the personal rights of RastafarI were routinely abrogated. Not only were members and supporters ostracized, discriminated against, and treated inhumanely, but they were killed across the Caribbean. While the exact numbers killed differ across sources, in Dominica (1975), a single police raid of a RastafarI commune resulted in 21 RastafarI from the House of Nyabinghi being killed (Phillips 2002; Sampson 1985). Similarly, in St. Lucia (1977), a police raid of a mountain RastafarI commune ended with one RastafarI member killed, an event memorialized in a local folk song. The bursts of police raids and acts of violence demonstrated the state's hard power against the RastafarI movement, which later turned to more systemic structural oppression.

Co-opted

RastafarI resistance, like other social protests, was critical counter-hegemonic responses to the Jamaican state. However, the state, composed of multiple political actors, possessed the ability to repress and also accommodate resistance movements to suit their interests. In the 1960s, student-led and working-class urban youth protested throughout the Jamaican capital of Kingston. The administration of the Michael Manley-led People's National Party (PNP) legitimized RastafarI upon acquiring state power, serving his political interest. Manley won the 1972 election by engaging radicalized youth through increased visibility of RastafarI aesthetics and discourse in the public domain (Meeks 2000, 121). Manley's appropriation of RastafarI symbols and selective embrace of the "radicalism and anti-hegemony of Rasta" ushered in a new period in the history of Jamaican politics (Sives 2002). Political co-optation took on a cultural element over time in response to global trends. For RastafarI, intentional resistance to colonial language resulted in the crafting of their own language called 'Iyaric' and the creation of reggae music as a self-made platform for self-representation (Barcant 2014). By the 1990s, with the global popularity of reggae music and Bob Marley, and the growth of the tourism industry, state-fueled targeting of RastafarI with hard power eased. The state utilized "romance tourism" and co-opted RastafarI images and music in national tourism campaigns to attract mass travelers with global marketing (Gordon 2020). The RastafarI movement continued to develop with regard to self-sufficiency in food production, communal farming, respect for nature, and food sovereignty. Growing in parallel was the global discourse and food sovereignty movement.

Reconciliation

As the food sovereignty concept and La Vía Campesina movement gained recognition worldwide, the formal state in the English-speaking Caribbean also embraced change to varying degrees. Similar to co-opting RastafarI imagery for tourism marketing, the state made steps to reconcile with the RastafarI to promote the state agenda and support improved public relations. For example, the Caribbean Community, an economic union of 18 member states (CARICOM), launched the first-ever commission in 2013 for financial reparations from European nations because of the detrimental social and economic impacts of chattel slavery and colonialism. These issues were previously discounted as RastafarI concerns (Atiles-Osoria 2018). In 2017, national efforts at rapprochement occurred, with financial compensation to RastafarI for the 1963 Coral Gardens Massacre by the Jamaican state (Gleaner Newspaper 2017). Two years later, a Cannabis Bill in St Kitts and Nevis recognized the use of marijuana 'ganja' as a sacrament in the RastafarI Faith. Despite these examples of reconciliation between the formal state and RastafarI movement, there have been few changes in formal, economic, and political institutions to support food sovereignty in the region.

Discussion

"The 'overstanding' and lived approach to language identified by Rastafari is the essence of reggae music. The ability of this music to relate to people across the globe ... the crux of its tactic of resistance" (Barcant 2014).

RastafarI as a Locally-Based Discourse and Praxis of Social and Environmental Sustainability

Over the early period, the RastafarI movement not only grew in numbers and became more broad-based, inclusive, and gender-sensitive, but it also embraced a stewardship ethic and politics of care in its belief system. Excluded from formal state support, they carved up communal space to produce food, protect nature, and reveal unappreciated connections between nature and human health.

It can be seen that institutions are not static, and internal change can be pragmatic. This can be seen in RastafarI dynamic repositioning of their principles toward Black supremacy and gender power relations. Since the 1970s, there has been increasing respect and roles for RastafarI women observed across various sects, customs, and social norms (Christensen 2014; Rowe 2012). Starting in the 1980s, the organization by RastafarI Daughters developed, and so did the emergence of hybridization in RastafarI multiethnic and international character. In 2000, one study estimated there were over one million self-defined RastafarI of multiple races located across the globe (Loadenthal 2013).

Institutional Interplay Between RastafarI and the Government of Jamaica

Institutional change arising from the influence of this social movement on formal institutions (the state) can be considered from multiple levels. Meadows (1999) identified the role of "intent" described as a deep leverage point. In describing the workings of "leverage points", Meadows (1999) showed that shifts in a social system influence substantial transformation in the social order by accommodating different societal goals and paradigms. The RastafarI shifted the colonial social order by creating a values-based, non-hierarchical social organization where individuals are tied together through common values and goals for equal rights and justice. Intent likely increased as racial, social, cultural, and historical commonalities broadened and membership numbers increased. In turn, the RastafarI social movement formed a collective intent that helped inflict systemic change within the formal institutions, as seen in the national reparation process.

These transformational moments occurred as this grassroots movement created momentum for change in various areas by articulating an alternative to the colonial viewpoint and developing new rules and norms. This shift from a rigid and hierarchical structure led to the development of a more participatory and engaging system. Other researchers have noted the importance of such institutional jockeying for enhanced social innovation. In discussing the role of agency in social-ecological transformation and change strategies, Westley et al. (2013) suggested that innovation occurs based on the degree to which institutional diversity is promoted and joint action fostered. Such engagement involves opportunities to interrogate existing arrangements, rules, and authority structures to accommodate needed change and foster new collaboration toward redefined goals. For example, accommodation by the state of RastafarI norms into the formal system introduces diverse values into the formal state apparatus, although outcomes involve differing degrees of impact and longevity.

Cross-scalar institutions may likely lead to continued change within formal institutions in Jamaica and the English-speaking Caribbean. For example, the push for human rights by the RastafarI was also supported by international institutions through global agreements (such as the adoption of the Universal Declaration of Human Rights in 1948). These likely served to enhance the legitimacy of changing formal national institutions toward a more rights-based policy agenda.

As seen in the co-opt phase (Figure 1), politicians incorporated RastafarI aesthetics during the 1970s election campaigns. This use of RastafarI messaging was an acknowledgment of the movement's attractiveness to radicalized urban youth, albeit for short-term political gains (Meeks 2000). Moreover, members continually challenge colonial holdover laws, agitating for change. One example can be found in a legal case in the judicial system seeking accommodation for the RastafarI sacrament of smoking marijuana. The ruling was described in the case of Ras Sankofa Maccabbee vs. In 2019, the ruling was described in the case of Sankofa Maccabbee vs. The Commissioner of Police (2017). The Eastern Caribbean Court of Justice ruled that the criminal prohibition of marijuana maintained colonial criminalizing practices in sanctions applied to cannabis use (while permitting medicinal, analytical, or research use). Such "Savings law" clauses were inserted by colonial authorities in the body of laws in the English-speaking Caribbean that held up colonial laws that pre-existed independence (Burnham 2005). While the legal system moved to accommodate the religious rights of the RastafarI, RastafarI's communal values of self-provisioning and environmental stewardship are ignored within state agricultural export-driven policies along with private leasehold tenures and high-input agriculture.

In the government's double postures with the RastafarI, the formal state likely gave rise to a phenomenon noted as an institutional void (Hajer 2003) and by Rahman et al. (2017) as an "inter-institutional gap". Rahman and colleagues (2017) describe how gaps underlying this pitfall exist because divergent rule-based interaction leads to resource degradation and conflict over resources when two institutions operate within the same resource system. In the case of RastafarI in Jamaica, we see that institutional changes such as co-optation and reconciliation with divergent goals, such as political gain and public relations, vs. genuine understanding and acceptance of communal values and food sovereignty can also lead to governance failures. Elements of this institutional gap have likely contributed to the lack of progress regarding food sovereignty practices and dialogue in the region.

Moreover, as Po et al. (2019) had questioned the premise that informal institutions are generally bounded by formal institutions, so too has the historical case of RastafarI movement shown that informal social movements can drive the formal state in reshaping its rhetorics and redressing historical injustice.

RastafarI and Their Intersectional Discourse of 'Ital' (Pure Food) and Livity (Natural Lifestyle) Showcasing the Primacy of Food to Equal Rights, Justice and Wellness of Human and Natural Life

At the start of their organizing, autonomous groups of RastafarI came together as part of a religious revivalist movement. By the 1970s, RastafarI had become engaged in self-provisioning that appeared to raise the importance and centrality of food to the movement and survival of members in Jamaica. In recognition of the importance of food, RastafarI devised new actions and social norms that challenged conventional neoliberalist principles of economic growth, increased privatization, and individual property rights with enhanced consumerism. Before the modern environmental groups raised focus on social, economic, and ecological impacts from high input, industrial, global agri-food systems (Sbicca 2016; McMichael 2011), RastafarI, to some degree, had articulated these concerns and devised new rules in response, suited to the Caribbean context.

The development of the RastafarI movement, with its new rules around local food and self-provision, directly challenges economic arguments for the global food trade of cheaper, long shelf life, caloric-dense but nutrient-poor processed foods, increasingly targeting low-income consumers and the developing world. Further, the movement challenges the conventional understanding of food systems that underestimate the social and cultural values of food. By recentering food in the RastafarI movement, RastafarI has institutionalized efforts to demonstrate the politics of food in locally specific ways. RastafarI explicitly confronts the intersectional issues of health, poverty, and food sovereignty, which are currently missing from the Nyéléni Declaration. RastafarI's concept of Ital is a useful decolonial framework where unhealthful processed foodstuffs should be eliminated from diets. This thinking aligns with the more recent foods classification, which is based on the degree and purpose of food processing: (i) unprocessed or minimally processed foods, (ii) processed culinary ingredients, (iii) processed foods, and (iv) ultra-processed foods and drink products (UPF) (Monterio et al. 2018).

Not only do the RastafarI explicitly reconnect food and health through principles such as Livity and Ital, but they also regard food as life-giving and not as a commodity. From this standpoint, RastafarI reframes solutions to hunger to align with aspirations for solidarity among communities rather than a zero-sum calculation. RastafarI's praxis aligns with critiques of the current food security discourse. As Rodríguez et al. (2017) describe, the mainstream food security discourse is non-political, has little to say on the commodification of food and concentration of economic power by transnational corporations in the hands of the few, and is the language of inter-governmental and multilateral organizations. Thus, RastafarI provides a measured response to what Burnett and Murphy (2014) suggest has been a limitation in the food sovereignty community. They suggest that the food sovereignty movement has not presented a clear position on the issue of trade. Discussions around socially responsible global value chains, such as fair-trade labeling, do not address the capitalistic narrative of global food systems. In contrast, the connections between health and self-provisioning of food that underpin RastafarI's Livity and Ital principles can guide the food sovereignty discourse to assert the collective rights of communities to indigenous knowledge practices, cosmologies, and sustainable relationships to the Earth (Chaifetz and Jagger 2014).

Conclusion

In tracing the historical development of the RastafarI in the Caribbean, this chapter contributes to the ongoing debates on food sovereignty as inherently political, social, and cultural and intersects with health, land resources, and racial and social justice in Jamaica. Our historical analysis shows that the place-based food sovereignty that RastafarI practices can necessarily be seen as a challenge to racial and class inequalities and hierarchies and power holders of the status quo. In the institutional interplay between the RastafarI movement and the state, we see how institutional diversity between formal and informal systems can change from state resistance using hard power to soft power, from RastafarI's resistance centered on racial justice and religious dogma to a social movement centered on a praxis of food self-provisioning and wellbeing in order to achieve social justice. However, the analysis also reveals that institutional diversity through co-optation and reconciliation on the surface will not lead to substantive transformation from dominant agri-food system discourse or oppression. In their development of a place-based food sovereignty discourse, RastafarI show that the environmental sustainability of food systems is interlinked with social sustainability; embedded within is the ability to make choices that enhance dignity and freedom. Gordillo and Méndez (2013) acknowledge this pivotal role of food in a rights-based framework because "without the right to food, neither life, nor human dignity, nor the enjoyment of other human rights can be assured." Future work should examine the historical and contemporary development of "rules-in-use" within the movement and how these processes differ across localities in the Caribbean. A better understanding of how these building blocks of rules support collective action can enhance natural resource governance outcomes and minimize future unintended inter-institutional pitfalls.

Acknowledgments

JYTP acknowledges funding from Research England (E3) through the Natural Resources Institute's Food and Nutrition Security Initiative (FaNSI).

References

Atiles-Osoria, J. 2018. Colonial state crimes and the CARICOM mobilization for reparation and justice. State Crime Journal 7(2): 349–368.

Bamikole, L.O. 2017. Livity as a dimension of identity in Rastafari thought: Implications for development in Africana Societies. Caribb Q. 63(4): 451–466.

Barcant, A. 2014. Language and power! Convergence Journal. Retrieved September 2, 2021, from http://convergencejournal.ca/volume-4.

Barnett, M. 2005. The many faces of Rasta: Doctrinal diversity within the Rastafari movement. Caribb Q 51(2): 67–78.

Baviskar, A. 2003. For a cultural politics of natural resources. Econ Polit Wkly, 5051–5055.

Beckford, G. 1972. Persistent poverty: Underdevelopment in plantation economies of the Third World. New York: Oxford University Press.

Beckford, C.D. Barker and S. Bailey. 2007. Adaptation, innovation and domestic food production in Jamaica: Some examples of survival strategies of small-scale farmers. Singap. J. Trop. Geogr. 28(3): 273–286.

Beckford, C.L. and D.R. Campbell. 2013. Food forests and home gardens: Their roles and functions in domestic food production and food security in the Caribbean. pp. 95–107. *In*: Domestic food production and food security in the Caribbean. Palgrave Macmillan, New York.

Béné, C., P. Oosterveer, L. Lamotte, I.D. Brouwer, S. de Haan, S.D. Prager et al. 2019. When food systems meet sustainability–Current narratives and implications for actions. World Dev. 113: 116–130.

Berkes, F. and C. Folke. 1998. Linking social and ecological systems for resilience and sustainability. Linking social and ecological systems: Management practices and social mechanisms for building resilience (pp. 1–25). Cambridge University Press, Cambridge.

Beuchelt, T. and D. Virchow. 2012. Food sovereignty or the human right to adequate food: Which concept serves better as international development policy for global hunger and poverty reduction? Agric. Human Values 29(2): 259–273.

Brissett, N.O. 2018. Education for social transformation (EST) in the Caribbean: A postcolonial perspective. Education Sciences 8(4): 197.

Burnett, K. and S. Murphy. 2014. What place for international trade in food sovereignty? J. Peasant. Stud. 41(6): 1065–1084.

Burnham, M.A. 2005. Saving constitutional rights from judicial scrutiny: The savings clause in the law of the Commonwealth Caribbean. Univ Miami Law Rev. 36(2/3): 249–269.

Campbell, H. 1987. Rasta and resistance: From Marcus Garvey to Walter Rodney. Africa World Press.

Cashmore, E.E. 1983. Rastaman: The Rastafarian movement in England. London: Unwin Paperbacks.

Chaifetz, A. and P. Jagger. 2014. 40 Years of dialogue on food sovereignty: A review and a look ahead. Glob Food Sec. 3: 85–91. https://doi.org/10.1016/j.gfs.2014.04.002.

Christensen, J. 2014. Rastafari reasoning and the RastaWoman: Gender constructions in the shaping of Rastafari Livity. Lanham: Lexington Books.

Cline, W.R. 2007. Global warming and agriculture: Impact estimates by country. Peterson Institute.

Dekeyser, K., L. Korsten and L. Fioramonti. 2018. Food sovereignty: Shifting debates on democratic food governance. Food Secur. 10(1): 223–233.

Edmonds, E.B. 2003. Rastafari: From outcasts to culture bearers. New York: Oxford University Press.

Eide, A. 1996. Human rights requirements to social and economic development. Food Policy 21(1): 23–39.

Fairbairn M. 2010. Framing resistance: International food regimes and the roots of food sovereignty. pp. 15–32. *In*: Hannah Wittman, Annette Desmarais and Nettie Wiebe (eds.). Food sovereignty: Reconnecting food, nature, and community, Publisher: Food First Books.

FAO. 2010. Low-income food-deficit countries (LIFDCs) - List for 2018. Retrieved from http://www.fao.org/countryprofiles/lifdc/en/.

Forsythe, D. 1980. West Indian culture through the prism of Rastafarianism. Caribb Q. 26(4): 62–81.

Girvan, N. 2005. W.A Lewis, The Plantation School and Dependence: An Interpretation. Soc. Econ. Stud. 54(3): 198–221.

Gjerset, H. 1994. First generation Rastafari in St. Eustatius: A case study in the Netherlands Antilles. Caribb Q 40(1): 64–77.

Gleaner Newspaper. 2017. We are sorry - Gov't apologises to Rastas for Coral Gardens incident. Retrieved from:http://jamaica-gleaner.com/article/lead-stories/20170405/we-are-sorry-govt-apologises-rastas-coral-gardens-incident.

Godfray, H.C.J., J.R. Beddington, I.R. Crute, L. Haddad, D. Lawrence, J.F. Muir et al. 2010. Food security: The challenge of feeding 9 billion people. Science 327(5967): 812–818.

Gordillo, G. and J. Méndez. 2013. Seguridad y soberanía alimentarias (documento base para discusión). FAO: Roma.

Gordon, A. 2020. Rastafarianism in Bullet Tree Falls, Belize: Exploring the effects of international trends. Societies 10(1): 24.

Hajer, M. 2003. Policy without polity? Policy analysis and the institutional void. Policy Sci. 36(2): 175–195.

Haugen, H.M. 2009. Food sovereignty–an appropriate approach to ensure the right to food? Nordic Journal of International Law 78(3): 263–292.

Hickey, G.M. and N. Unwin. 2020. Addressing the triple burden of malnutrition in the time of COVID-19 and climate change in Small Island Developing States: What role for improved local food production?. Food Secur. 12(4): 831–835.

Homiak, J.P. 2005. Ethiopia arisen: Discovering Rastafari. AnthroNotes 26(2): 10–18.

Jayaprakash, L.G. and G.M. Hickey. 2019. Mistaking the map for the territory: What does the history of Bannerghatta National Park, India, tell us about the study of institutions? Soc. Nat. Resour. 32(12): 1433–1450.

La Vía Campesina (LVC). 2019. 2019 Annual Report. https://viacampesina.org/en/la-via-campesina-2019-annual-report/ (Accessed 4th November 2020).

La Vía Campesina (LVC). 2020. The international peasants' voice! Globalising hope, globalising the struggle! https://viacampesina.org/en/international-peasants-voice/#:~:text=La%20Via%20Campesina%20is%20an,workers%20from%20around%20the%20world. (Accessed 4th November 2020).

Loadenthal, M. 2013. Jah people: The cultural hybridity of white Rastafarians. Glocalism: Journal of Culture, Politics and Innovation 1(1): 1–21.

Lowitt, K., G.M. Hickey, W. Ganpat and L. Phillip. 2015. Linking communities of practice with value chain development in smallholder farming systems. World Dev. 74: 363–373.

Lowitt, K., A. Saint Ville, C. Keddy, L.E. Phillip and G.M. Hickey. 2016. Challenges and opportunities for more integrated regional food security policy in the Caribbean Community. Reg. Stud. Reg. Sci. 3(1): 706–716.

Mckeon, L. 2017. The true story of RastafarI. Retrieved from: https://www.nybooks.com/daily/2017/01/06/the-true-story-of-rastafari/.

McMichael, P. 2011. Food system sustainability: Questions of environmental governance in the new world (dis) order. Glob Environ Change 21(3): 804–812.

Meadows, D.H. 1999. Leverage points: Places to intervene in a system. Sustainability Institute, VT.

Meadows, D.H., J. Randers and D.L. Meadows. 2009. Grenzen des wachstums: das 30-jahre-update. Stuttgart: Hirzel.

Meeks, B. 2000. Narratives of resistance: Jamaica, Trinidad, The Caribbean. Kingston, Jamaica: UWI Press.

Mintz, S.W. 1985. From plantations to peasantries in the Caribbean. pp. 127–153. *In*: S. Mintz and S. Price (eds.). Caribbean contours. Baltimore and London: Johns Hopkins University Press.

Monteiro, C.A., G. Cannon, J.C. Moubarac, R.B. Levy, M.L.C. Louzada and P.C. Jaime. 2018. The UN decade of nutrition, the NOVA food classification and the trouble with ultra-processing. Public Health Nutr. 21(1): 5–17.

Nair, P.R. 1993. An introduction to agroforestry. Springer Science and Business Media.

North, D. 1991. Institutions. J. Econ. Perspect. 5(1): 92.

Nyéléni. 2007. The Declaration of "Nyéléni". Retrieved from https://nyeleni.org/spip.php?article290.

Ostrom, E. 2005. Doing institutional analysis digging deeper than markets and hierarchies. pp. 819–848. *In*: Handbook of new institutional economics. Springer, Boston, MA.

Ostrom, E., J. Walker and R. Gardner. 1994. Rules, games, rules and common pool resources. Cambridge,MA: The MIT Press.

Ostrom, V. 1999. Polycentricity (part 1). Polycentricity and local public economies, 52–74.

Pantin, D. 1980. The plantation economy model and the Caribbean. The IDS Bulletin 12(1): 17–23.

Patel, R. 2009. Food Sovereignty. J Peasant Stud, 36, 3: 663–706.

Phillips, D.E. 2002. The defunct Dominica defense force and two attempted coups on the nature island. Caribb Stud, 52–81.

Po, J.Y.T., A. Saint Ville, H.M.T. Rahman and G.M. Hickey. 2019. On institutional diversity and interplay in natural resource governance. Soc. Nat. Resour. 32(12): 1333–1343.

Rahman, H.M., A.S. Saint Ville, A.M. Song, J.Y. Po, E. Berthet, J.R. Brammer et al. 2017. A framework for analysing institutional gaps in natural resource governance. International Journal of the Commons, 11(2): 823–853.

Ras Sankofa Maccabbee v The Commissioner of Police, (Eastern Caribbean Supreme Court, 2017. https://www.eccourts.org/ras-sankofa-maccabbee-v-the-commissioner-of-police/.

Richardson, B.C. 1992. The Caribbean in the wider world, 1492–1992: A regional geography. Cambridge University Press, Cambridge. doi:10.1017/cbo9780511560057.

Rodríguez, E.B., J.J. Heredia and G.K. Velasco. 2017. Food security in the European Union, Latin America and the Caribbean: The Cases of Cuba and Spain. EU-LAC Foundation.

Rowe, M. 2012. The woman in RastafarI. *In*: Barnett, M. (ed.). (2014). Rastafari in the new millennium: A Rastafari reader. Syracuse University Press.

Sampson, G.E. 1985. Religion and justice: Some reflections on the Rastafari movement. Phylon (1960-), 46(4): 286–291.

Saint Ville, A.S., G.M. Hickey and L.E. Phillip. 2015. Addressing food and nutrition insecurity in the Caribbean through domestic smallholder farming system innovation. Reg Environ Change 15(7): 1325–1339.

Sbicca, J. 2016. Food sovereignty or bust: Transforming the agrifood system is a must. pp. 320–332. *In*: Emergent possibilities for global sustainability. Routledge.

Sives, A. 2002. Changing patrons, from politician to drug don: Clientelism in downtown Kingston, Jamaica. Lat Am Perspect 29(5): 66–89.

Timms, B. 2008. Development theory and domestic agriculture in the Caribbean: Recurring crises and missed opportunities. Caribb. Geogr. 15(2): 101–117.

Toussaint, M., P. Cabanelas and P. Muñoz-Dueñas. 2021. Social sustainability in the food value chain: what is and how to adopt an integrative approach?. Qual Quant, 1–24.

Warsh, M.A. 2014. A political ecology in the early Spanish Caribbean. William Mary Q, 71(4): 517–548.

Watts, D. 1990. The West Indies: Patterns of development, culture, and environmental change since 1492. Cambridge University Press, Cambridge.

Whatmore, S. and L. Thorne. 1997. Nourishing networks: Alternative geographies of food in Goodman D and Watts M eds Globalising food: agrarian questions and global restructuring. New York, 287–30.

Weis, T. 2003. Agrarian decline and breadbasket dependence in the Caribbean: Confronting illusions of Inevitability. Labour Cap. Soc., 174–199.

Westley, F.R., O. Tjornbo, L. Schultz, P. Olsson, C. Folke, B. Crona et al. 2013. A theory of transformative agency in linked social-ecological systems. Ecol. Soc., 18(3).

Winter, M. 2005. Geographies of food: agro-food geographies-food, nature, farmers and agency. Prog. Hum. Geogr. 29(5): 609–617. https://doi.org/10.1191/0309132505ph571pr.

Wittman, H., A. Desmarais and N. Wiebe. 2010. The origins and potential of food sovereignty. Food sovereignty: Reconnecting Food, Nature and Community, 1–14.

7

Diversity in Institutional Strategies for Distributed Renewable Energy Generation
How American States are Designing Net Metering Policy

Saba Siddiki,[1,*] *Kathleen Smith*[2] and *Chris Koski*[3]

Introduction

Scholars of environmental governance have long recognized the importance of institutional diversity (Ostrom 2005). Institutional diversity means variety in the strategies, norms, and rules (e.g., policies or social expectations) used to govern behavior (Crawford and Ostrom 1995). In the context of environmental governance, this diversity is manifest in variations in the designs of institutions used to resolve environmental dilemmas. The call for institutional diversity in environmental governance stems from an appreciation of the unique contexts in which environmental dilemmas embed, reflected in how the qualities of environmental goods differ across contexts as well as how characteristics of users of environmental goods differ across contexts. As communities worldwide grapple with enduring, evolving, and emergent environmental dilemmas linked to global climate change, increased prevalence and salience of institutional diversity appear likely. Evolving kinds and qualities of environmental dilemmas, coupled with their variable presentation across communities, are likely to continue to elicit context-specific strategies.

[1] Syracuse University, 426, Eggers Hall, Syracuse, NY 13244.

[2] 426 Eggers Hall, Syracuse, NY 13244.
Email: ksmith71@syr.edu

[3] Reed College, 3203 SE, Woodstock Blvd, Portland, OR 97202.
Email: ckoski@reed.edu

* Corresponding author: ssiddiki@syr.edu

In the United States, institutional diversity in environmental governance is accommodated by delegating policymaking responsibility and authority to the states. Across a wide array of environmental domains, including wetland protection, management of emissions from the transportation sector, climate action planning, and others, the federal government has afforded single or multiple states the right to design policies or tailor implementation of federal environmental laws to suit unique social and environmental contexts (Arnold 2015; Engel 2006; Karapin 2016). In recent decades, states have leveraged this authority to develop distinctive and innovative institutions to encourage increased reliance on renewable energy sources. This comes with the recognition that increased reliance on renewable energy sources in major industries and sectors can reduce greenhouse gas emissions, one of the major contributors to global climate change.

In the electricity sector, net metering policy has emerged as one prominent institutional strategy to increase reliance on renewable energy. Net metering enables households, commercial, and/or industrial entities to sell back to utilities unused electricity they produce from renewable energy infrastructure installed on their properties. Over the last twenty years, American states have widely adopted net metering policies. Different states are focused on distinctive arrays of targets, instruments, and incentives within the context of their policies' designs. In this chapter, we provide a descriptive account of policies that states have adopted to encourage the use of net metering toward the broader objective of promoting environmental sustainability. In doing so, we leverage the concept of "robustness" to offer a more expansive characterization of variation in the design of states' net metering policies. Accordant with dimensions of robustness, state policies are characterized in terms of diversity, modularity, and redundancy.

We begin the chapter with an elaborate discussion of institutional analysis, policy design, robustness, and how we integrate these within our analysis. This is followed by an elaborated discussion of net metering, the methods through which we assess variation in the designs of state net metering policies, the results from our assessment, and a brief discussion summarizing key insights gleaned from our analysis.

Analyzing Institutions

Institutional analysis involves the study of strategies, norms, and rules that govern behavior (Crawford and Ostrom 1995). Within institutional scholarship, strategies, norms, and rules are all considered types of "institutions", which convey with varying degrees of prescriptive force what actors are required, allowed, or forbidden to do or constitute features of institutionally governed systems (Frantz and Siddiki 2021). Among the central foci of institutional scholarship is understanding the institutional formation, design, evolution, and impacts (Ghorbani and Bravo 2016; Siddiki et al. 2019). Institutional scholarship features studies that explore these topics in a variety of domains, though a sizable portion has been dedicated to evaluating institutional dynamics relating to environmental governance. By exploring environmental governance using theory and methods for supporting institutional analysis, scholars have sought to uncover the combinations of institutions that communities use to overcome collective action dilemmas relating to shared usage of environmental goods (Ostrom 1990), the array of informal institutions that communities use in natural resource management (Basurto 2005), varying institutional arrangements through which governments seek to develop, distribute, and protect depletable natural resources, among a host of other topics (Blomquist et al. 2001). Explicitly or implicitly relayed in this research is an appreciation of institutional diversity, indicating that communities pursue distinctive arrays of institutional strategies to address environmental dilemmas.

To aid in the investigation of institutional design, which is foundational to the study of institutional diversity, scholars have drawn on various theories and frameworks that offer guidance on generalizable features of institutions along which institutions can be characterized. Among the most prominent of these used by institutional scholars is common pool resource theory, which offers guidance on classifying institutions according to their functional properties and analyzing how institutions with different functional properties link together to support successful common pool resource governance. The affiliated institutional analysis and development framework offers even more generalized guidance on studying the development, form, and outcomes of formal and informal institutions that govern behavior. Scholars particularly interested in evaluating the design of formal institutions manifesting as public policy have relied on theories of policy design which, like the approaches described above, identify common elements of policy design along which individual policies can be characterized and along which multiple policies can be compared (Schneider and Ingram 1988). For example, Schneider and Ingram (1997) argue that policies commonly convey information about policy targets (information about to whom a policy applies), policy instruments (information about the specific tools through which policies will compel behavior), and in connection with policy instruments and policy incentives (behavioral incentives).

Institutional scholars exploring environmental governance have consistently borrowed concepts from research in the natural sciences that foster an understanding of system-level dynamics; for example, ecological research focused on ecosystem resilience and robustness (Capano and Woo 2017; Folke et al. 2010). The concepts of resilience and robustness are usually used to characterize states of more or less dynamic systems, where dynamism is assumed to result from typical flux in system features or could result from unexpected shocks. While conceptually distinct, both robustness and resilience essentially refer to a system's ability to sustain function amidst various types of change. Three features of systems are expected to contribute to their robustness or ability to sustain: diversity, modularity, and redundancy. In general terms, diversity references variety in system features; modularity references functional independence among system features, and redundancy references replication or overlap in system features (Kharazzi et al. 2020). Both separately and in combination, diversity, modularity, and redundancy make a system less vulnerable to collapse, should one system feature fail or falter.

As reiterated above, robustness is presumed to be an important quality of dynamic, and even complex and uncertain, systems insofar as it contributes to their sustained functionality. In recent years, social scientists have also alluded to the importance of having robust institutions to govern socio-environmental systems that are dynamic, complex, and uncertain. An assumption underlying this translation from the systemic to governance perspective is that robust institutions will sustain effectiveness even as the systems they govern, or knowledge relating thereto, changes. Toward this end, institutional scholars have adapted conceptualizations of ecological system robustness and specific indicators thereof and suggested how they might operationalize in public policy design. For example, Anderies and Janssen (2017) adapt the three indicators of robustness reviewed above—diversity, modularity, and redundancy—to the policy domain. They define each indicator as the following, where each could be understood as a characteristic of a public policy: (i) diversity—having a multiplicity of different types of regulatory mechanisms; (ii) redundancy—having many regulatory mechanisms perform similar functions; and (iii) modularity—having some regulatory mechanisms with only limited connectivity with others. Characterization of modularity essentially accommodates understanding of groupings of independent regulatory mechanisms. Characterizing robustness in policy design along these different indicators enables various insights into how regulatory mechanisms are embodied within policy. More generally, it offers a conceptual basis for exploring different forms of variation—diversity in the wider sense—in institutional design.

Although capturing robustness in the design of public policy may imply a sense of rigidity, this need not be the case. As we will describe later, the different program characteristics used in our study of net metering policy have a degree of flexibility within them that lawmakers can harness to adjust the policy to new obstacles or opportunities. For example, caps placed on the size of the net metering systems installed by utility customers, which restrict the amount of energy that can be produced and sold, can be increased to encourage the production of more renewable energy or decreased to restrict it. This flexibility allows net metering policies to be robust yet adjust to instances, such as an increase or decrease in consumer interest. This concept can be observed in the ever-evolving state net metering policies that are regularly adjusted by lawmakers as new technologies develop or states desire to enact more environmentally friendly policies.

In the remainder of this chapter, we describe how we use policy design scholarship in connection with the concept of robustness and related indicators of diversity, redundancy, and modularity toward a study of variation in the design of policies American states use to support net metering.

Net Metering

Net metering policies encourage the installation of distributed electricity generation infrastructure in establishing a mechanism through which utility customers can sell back to utilities at a specified compensation rate for unused electricity generated from this infrastructure (National Council of State Legislatures 2017). Common types of distributed electricity infrastructure include photovoltaic solar arrays, wind turbines, and small-scale geothermal plants. Compared to commercial electricity generators that can only be installed in certain places and at large scales to maintain profitability, distributed electricity energy generation infrastructures can be connected to smaller buildings (e.g., residential units) and operate at various scales. Distributed electricity energy generation infrastructure is, thus, deemed more flexible and easily integrated with existing structures than more centralized systems.

Distributed electricity generation has increased in prevalence in the last decade in the United States in response to technological advancements, rising energy costs, and concerns about environmental pollution from the combustion of fossil fuels. The net metering policy provides a formal institutional strategy through which to encourage it. Currently, 45 states have adopted a net metering policy. There were notable waves in net metering policy enactment. Between 1997 and 2001, 15 states enacted net metering policies. Between 2007 and 2009, an additional seven states followed suit.

States' net metering policies commonly address policy targets (who is eligible to participate in net metering programs), instruments (mechanisms for controlling renewable energy capacity), and incentives (financial and crediting incentives for encouraging participation in net metering programs). However, states differ substantially in how they address each of the policy design elements, offering an opportunity to investigate how different states pursue different institutional strategies in a common policy domain.

In the following section, we describe how we captured information on states' net metering policies in terms of policy targets, instruments, and incentives and how this information was used to conduct a descriptive assessment of institutional diversity, modularity, and redundancy.

Methods

Our descriptive summary of state net metering policy was motivated by the objective of capturing how states operationalize the following three policy design elements: policy targets, policy instruments, and policy incentives. These elements are identified as theoretically salient features of policy design (Schneider and Ingram 1997). To support our identification of operational measures of each of these elements in the context of net metering policy, we identified types of information commonly conveyed in states' net metering policies. We chose measures that are recognized as being of practical import and among which policies exhibit variation that could be exploited as part of our descriptive analysis. We reviewed secondary policy resources containing information about the net metering domain and actual net metering policies as part of this effort. State policies were collected from the Database of State Incentives for Renewables & Efficiency (DSIRE). Our analysis reports on policies states had in place in 2017. We collected the following operational measures of policy targets, instruments, and incentives for each state's policy.

Policy Targets

Types of utilities allowed to participate in a state's net metering program. Across policies, we observe that states allow one or more of the following types of utilities to participate in their net metering program: investor-owned, municipal, or cooperative.

Thus, for policy targets, we identified:

- Whether a state allows investor-owned utilities to participate in its net metering program (0/1).
- Whether a state allows municipally-owned utilities to participate in its net metering program (0/1).
- Whether a state allows cooperatively owned utilities to participate in its net metering program (0/1).

Policy Instruments

The identification of distributed electricity generation system caps, overall program caps, and allowance for aggregated metering within a state's net metering program is crucial. System caps identify the maximum size of net metering customers' electricity generation systems. Program size caps limit the total amount of electricity that utilities can buy back from net metering customers. Aggregate metering allows customers to aggregate their electricity use across meters or buildings when multiple meters or buildings are drawing from the same distributed electricity infrastructure. Thus, for policy instruments, we identified:

- Whether a state identifies within its net metering policy a system cap (0/1).
- Whether a state identified within its net meting policy a program size cap (0/1).
- Whether a state includes an allowance for aggregate metering in its net metering policy (0/1).

Policy Incentives

Compensation rates for electricity sold back to utilities by customers participating in a state's net metering program, the absence of a disincentive to rollover "renewable energy credits" obtained through participation in a net metering program, and whether customers or utilities are owners of these renewable energy credits is considered. Compensation under net metering programs is conducted by awarding credits for the amount of electricity transacted. States differ in how they value credits; for example, based on whether they compensate consumers for their electricity at a wholesale (lesser) or retail (greater) rate, which is then reflected in the value of the credits. States also differ in whether they allow customers to "roll over" credits across billing cycles. Finally, states differ in which entity they deem to own renewable energy credits. In some states, customers are considered the owner of credits. In other states, utilities are the owners of credits; they earn credit for every unit of electricity their customers produce from renewable energy infrastructure.

Thus, for policy incentives, we identified:

- Whether a state compensates net metering customers for excess electricity at a retail rate rather than another compensation rate such as wholesale (0/1).
- Whether a state does not disincentivize the rollover of renewable energy credits across billing cycles (0/1).
- Whether a state identifies the net metering customer, rather than the utility, as the owner of renewable energy credits (0/1).

Once we constructed operational measures for policy targets, instruments, and incentives, we then used these measures to determine, by state, institutional diversity, institutional redundancy, and institutional modularity. Institutional diversity, institutional redundancy, and institutional modularity each represent a different indicator of institutional robustness. Descriptions of how each of these was determined are provided below. We compared information collected on these three indicators of institutional robustness to glean a descriptive understanding of the state net metering policy landscape.

Institutional Diversity

Our measurement of institutional diversity captures how many different types of policy elements are addressed by a state in its net metering policy; that is, whether the state provides any guidance on policy targets, policy instruments, or policy incentives, which we assess through the aforementioned operational measures. Though we have multiple measures per policy design element, we give states a value of 1 if any operational measures associated with a policy design element are addressed. Thus, for institutional diversity, a state can score from 0 (reflecting that the policy does not address policy targets, instruments, or incentives based on our operational measures of such) to 3 (reflecting that the policy has addressed all three policy elements).

Institutional Redundancy

Our institutional redundancy score is designed to capture if/how a state addresses each operational measure we use for the three policy design elements, supporting an understanding of the extent to which an element is addressed. As indicated in our description of operational measures, we are interested in both whether and to what extent a state addresses an element measure. In combination, we are able to glean insights about which policy design elements are more or less thoroughly addressed by a state. For institutional redundancy, we assign three values for each state, reflecting how many policy instrument measures are addressed (0–3), how many policy incentives measures are addressed (0–3), and how many different types of policy targets can participate in a state's net metering program (0–3).

Institutional Modularity

Institutional modularity captures elements that represent independent mechanisms in policy design. Individual measures for policy targets, policy instruments, and policy incentives represent distinct ways of influencing net metering adoption and use. For each of the three policy targets, we identify independent regulatory mechanisms because they specifically address different groups. Each of the three policy instruments we identify is independent because they address unique ways of approaching limits to net metering systems. Our three incentive measures represent two distinct mechanisms of compensation: rate of compensation for net metered electricity and control over renewable energy credit (who owns these credits and what they can do with them). Our institutional modularity score builds on the institutional redundancy score by capturing how many operational measures, according to the different policy design elements, are addressed in a state's net metering policy, either through any reference or through the assignation of particular values. For institutional modularity, we sum each state's institutional redundancy values (though we combine REC ownership and rollover given the dependence of these items for a maximum score of 2 for incentives) to arrive at one value per state, indicating total measures across elements addressed.

Results

We summarize results from our descriptive assessment of states' net metering policies around our three indicators of robustness—institutional diversity, institutional redundancy, and institutional modularity. Diversity scores are reported in Table 1, redundancy scores in Table 2, and modularity scores in Table 3.

Assessing states' scores on institutional diversity, which tells us how many of the three policy design elements (instruments, incentives, and targets) states are addressing in their policies, we find that most states are addressing all three. Of the 45 states that currently have a net metering policy, 39 states address policy targets, policy instruments, and policy incentives in some way. Six states address only two policy design elements, five only address policy instruments and policy targets, and one only addresses policy incentives and policy targets. Institutional redundancy scores tell us which operational measures correspond to the three design elements states address. Regarding policy instruments, nearly all states specify a system cap in the net metering policies, almost half identify a program cap, and about 40 percent allow for aggregate metering. For policy incentives, about two-thirds of states specify a retail compensation rate for net metering customers, approximately one-third do not disincentivize the rollover of renewable energy credits, and about half of states award the ownership of renewable energy credits to the customer. The institutional modularity scores, which reflect the sum of each state's independent regulatory mechanisms to address net metering to arrive at one value per state indicating total measures across elements addressed, reveal more generally the diversity in the design of states' net metering policies. Modularity scores range from 2–8 (on a possible range of 0–8), suggesting that some states are addressing multiple policy design elements through a single operational measure per element while others are addressing multiple policy design elements through multiple measures.

Discussion

Evaluating the design of American states' 2017 net metering policies affords an understanding of the various strategies that different jurisdictions are pursuing to support increased reliance on renewable energy sources. Increased reliance on renewable energy sources and concomitant reductions in greenhouse gas emissions are viewed as important strategies for encouraging environmental sustainability goals. We draw three main insights from our assessment of states' net metering policies.

First, different assessments of institutional variation yield distinctive, albeit complementary, insights about the different strategies through which states are

Table 1: Institutional diversity by state for net metering policy.

State	Policy Instruments	Policy Incentives	Policy Targets	Total Elements Addressed
Alabama	n/a	n/a	n/a	n/a
Alaska	1	1	1	3
Arizona	1	1	1	3
Arkansas	1	1	1	3
California	1	1	1	3
Colorado	1	1	1	3
Connecticut	1	1	1	3
Delaware	1	1	1	3
Florida	1	1	1	3
Georgia	1	0	1	2
Hawaii	1	1	1	3
Idaho	n/a	n/a	n/a	n/a
Illinois	1	1	1	3
Indiana	1	1	1	3
Iowa	1	1	1	3
Kansas	1	0	1	2
Kentucky	1	1	1	3
Louisiana	1	1	1	3
Maine	1	1	1	3
Maryland	1	1	1	3
Massachusetts	1	1	1	3
Michigan	1	1	1	3
Minnesota	1	1	1	3
Mississippi	1	1	1	3
Missouri	1	0	1	2
Montana	1	1	1	3
Nebraska	1	1	1	3
Nevada	1	1	1	3
New Hampshire	1	1	1	3
New Jersey	1	1	1	3
New Mexico	1	1	1	3
New York	1	1	1	3
North Carolina	1	1	1	3
North Dakota	1	1	1	3
Ohio	0	1	1	2
Oklahoma	1	1	1	3

Table 1 contd. ...

...Table 1 contd.

State	Policy Instruments	Policy Incentives	Policy Targets	Total Elements Addressed
Oregon	1	1	1	3
Pennsylvania	1	1	1	3
Rhode Island	1	0	1	2
South Carolina	1	1	1	3
South Dakota	n/a	n/a	n/a	n/a
Tennessee	n/a	n/a	n/a	n/a
Texas	n/a	n/a	n/a	n/a
Utah	1	1	1	3
Vermont	1	1	1	3
Virginia	1	1	1	3
Washington	1	1	1	3
West Virginia	1	1	1	3
Wisconsin	1	1	1	3
Wyoming	1	0	1	2

encouraging net metering. Leveraging the three indicators of robustness in the descriptive assessment of policy design allows us to discern variation in states' net metering policies through different lenses. Observing institutional diversity scores does not reveal the true variation in states' policies, as most states appear to be addressing multiple policy design elements with some operational measure. Institutional modularity scores offer slightly richer insights into policy variation by providing a sense of the number of independent operational measures across policy design elements states are addressing. Richer insights are revealed through institutional redundancy scores, which suggest how many operational measures by element states are addressed in their net metering policies, which also convey information about the combinations of measures reflected in states' policies.

Second, beyond validating its practical utility, our analysis broadly supports using the concept of robustness within institutional analysis (Brady 2020). The concept of institutional diversity has long interested institutional scholars (Ostrom 2005). Leveraging the concept of robustness can aid institutional scholars in the conceptualization of institutional diversity, or variation, in more multi-dimensional terms, ultimately toward supporting more nuanced characterizations of institutional design. Our particular analysis also shows how robustness can be conceptualized and operationalized with reference to approaches for characterizing policy design, given that our specific analysis focuses on such. We find value in structuring indicators of robustness—institutional diversity, institutional redundancy, and institutional modularity—around policy instruments, policy incentives, and policy targets (Schneider and Ingram 1997).

Third, by using operational measures that, in some cases, account for the presence of certain types of information alongside measures that capture characteristics, or

Table 2: Institutional redundancy by state for net metering policy.

State	Policy Instruments				Policy Incentives				Policy Targets			
	System Cap	Program Cap	Meter Aggregation	Total Measures Addressed by Element	Excess Electricity Payment Rate	Absence of Rollover Disincentive	Ownership of RECs	Total Measures Addressed by Element	Investor-Owned Utility	Municipal Utility	Electric Co-operative	Total Measures Addressed by Element
Alabama	n/a	n/a	n/a	n/a	n/a	n/a	n/a	n/a	n/a	n/a	n/a	n/a
Alaska	1	1	0	2	0	1	0	1	1	1	1	3
Arizona	1	0	0	1	0	0	1	1	1	0	1	2
Arkansas	1	0	1	2	1	0	1	2	1	0	1	2
California	1	0	1	2	1	0	1	2	1	1	1	3
Colorado	1	0	1	2	1	0	1	2	1	1	1	3
Connecticut	1	0	0	1	1	0	1	2	1	0	0	1
Delaware	1	1	1	3	1	0	1	2	1	1	1	3
Florida	1	0	0	1	1	0	1	2	1	0	0	1
Georgia	1	1	0	2	0	0	0	0	1	1	1	3
Hawaii	1	1	0	2	1	0	0	1	1	1	1	3
Idaho	n/a	n/a	n/a	n/a	n/a	n/a	n/a	n/a	n/a	n/a	n/a	n/a
Illinois	1	1	0	2	1	0	1	2	1	0	0	1
Indiana	1	1	0	2	1	1	0	2	1	0	0	1
Iowa	1	0	0	1	1	0	0	1	1	0	0	1
Kansas	1	1	0	2	0	0	0	0	1	0	0	1
Kentucky	1	1	0	2	1	1	1	3	1	0	1	2
Louisiana	1	1	0	2	1	0	0	1	1	1	1	3
Maine	1	0	1	2	1	0	0	1	1	1	1	3
Maryland	1	1	1	3	1	0	1	2	1	1	1	3
Massachusetts	1	1	0	2	0	1	1	2	1	0	0	1
Michigan	1	1	0	2	1	1	1	3	1	1	1	3

State	Policy Instruments				Policy Incentives				Policy Targets			
	System Cap	Program Cap	Meter Aggregation	Total Measures Addressed by Element	Excess Electricity Payment Rate	Absence of Rollover Disincentive	Ownership of RECs	Total Measures Addressed by Element	Investor-Owned Utility	Municipal Utility	Electric Co-operative	Total Measures Addressed by Element
Minnesota	1	0	1	2	1	1	1	3	1	1	1	3
Mississippi	1	1	0	2	0	1	1	2	1	1	1	3
Missouri	1	1	0	2	0	0	0	0	1	1	1	3
Montana	1	0	0	1	1	0	0	1	1	0	0	1
Nebraska	1	1	0	2	0	1	1	2	1	1	1	3
Nevada	1	0	1	2	1	0	1	2	1	0	0	1
New Hampshire	1	0	0	1	0	1	0	1	1	1	1	3
New Jersey	0	0	1	1	1	0	1	2	1	0	0	1
New Mexico	1	0	0	1	0	1	0	1	1	0	1	2
New York	1	0	1	2	1	0	0	1	1	0	0	1
North Carolina	1	0	0	1	1	0	0	1	1	0	0	1
North Dakota	1	0	0	1	0	1	0	1	1	0	0	1
Ohio	0	0	0	0	0	1	0	1	1	0	0	1
Oklahoma	1	0	0	1	0	1	0	1	1	0	1	2
Oregon	1	1	1	3	1	0	1	2	1	1	1	3
Pennsylvania	1	0	1	2	1	0	1	2	1	0	0	1
Rhode Island	1	1	1	3	0	0	0	0	1	0	0	1
South Carolina	1	1	0	2	1	0	0	1	1	1	0	2
South Dakota	n/a	n/a	n/a	n/a	n/a	n/a	n/a	n/a	n/a	n/a	n/a	n/a
Tennessee	n/a	n/a	n/a	n/a	n/a	n/a	n/a	n/a	n/a	n/a	n/a	n/a

Table 2 contd. ...

...Table 2 contd.

State	Policy Instruments				Policy Incentives				Policy Targets			
	System Cap	Program Cap	Meter Aggregation	Total Measures Addressed by Element	Excess Electricity Payment Rate	Absence of Rollover Disincentive	Ownership of RECs	Total Measures Addressed by Element	Investor-Owned Utility	Municipal Utility	Electric Co-operative	Total Measures Addressed by Element
Texas	n/a	n/a	n/a	n/a	n/a	n/a	n/a	n/a	n/a	n/a	n/a	n/a
Utah	1	1	1	3	1	0	1	2	1	0	1	2
Vermont	1	0	1	2	0	0	1	1	1	1	1	3
Virginia	1	1	1	3	1	0	1	2	1	0	1	2
Washington	1	1	1	3	1	0	1	2	1	1	1	3
West Virginia	1	1	1	3	1	1	0	2	1	1	1	3
Wisconsin	1	0	0	1	1	1	0	2	1	1	0	2
Wyoming	1	0	0	1	0	0	0	0	1	0	1	2
Total	43	22	18	44	29	15	23	40	45	21	27	45

Table 3: Institutional modularity by the state for net metering policy.

State	Instruments	Incentives	Targets	Total Elements Addressed	Total Measures Across Elements Addressed
Alabama	n/a	n/a	n/a	n/a	n/a
Alaska	2	1	3	3	6
Arizona	1	1	2	3	4
Arkansas	2	2	2	3	6
California	2	2	3	3	7
Colorado	2	2	3	3	7
Connecticut	1	2	1	3	4
Delaware	3	2	3	3	8
Florida	1	2	1	3	4
Georgia	2	0	3	2	5
Hawaii	2	1	3	3	6
Idaho	n/a	n/a	n/a	n/a	n/a
Illinois	2	2	1	3	5
Indiana	2	2	1	3	5
Iowa	1	1	1	3	3
Kansas	2	0	1	2	3
Kentucky	2	2	2	3	6
Louisiana	2	1	3	3	6
Maine	2	1	3	3	6
Maryland	3	2	3	3	8
Massachusetts	2	1	1	3	4
Michigan	2	2	3	3	7
Minnesota	2	2	3	3	7
Mississippi	2	1	3	3	6
Missouri	2	0	3	2	5
Montana	1	1	1	3	3
Nebraska	2	1	3	3	6
Nevada	2	2	1	3	5
New Hampshire	1	1	3	3	5
New Jersey	1	2	1	2	4
New Mexico	1	1	2	3	4
New York	2	1	1	3	4
North Carolina	1	1	1	3	3
North Dakota	1	1	1	3	3
Ohio	0	1	1	2	2

Table 3 contd. ...

...Table 3 contd.

State	Instruments	Incentives	Targets	Total Elements Addressed	Total Measures Across Elements Addressed
Oklahoma	1	1	2	3	4
Oregon	3	2	3	3	8
Pennsylvania	2	2	1	3	5
Rhode Island	3	0	1	2	4
South Carolina	2	1	2	3	5
South Dakota	n/a	n/a	n/a	n/a	n/a
Tennessee	n/a	n/a	n/a	n/a	n/a
Texas	n/a	n/a	n/a	n/a	n/a
Utah	3	2	2	3	7
Vermont	2	1	3	3	6
Virginia	3	2	2	3	7
Washington	3	2	3	3	8
West Virginia	3	2	3	3	8
Wisconsin	1	2	2	3	5
Wyoming	1	0	2	2	3

values, associated with this information, we are able to draw insights not only about what states deem most critical to address in policy design but also how they are distributing policy benefits (i.e., to whom policy incentives are being directed). The absence of disincentives for the rollover of renewable energy credits across billing cycles and assigning ownership of credits to utility customers favors customers over utilities. Though we treat our operational measures of each policy design element additively (i.e., sum the number of operational measures addressed through a state's policy design), assessing the values of the operational measures gives us a more nuanced view of how these policies benefit and burden different target groups. From observing values for the operational measures of policy incentives, we see at a high level that policy benefits are more heavily accruing to utility customers than utilities. Subsequent analysis of data collected for this research will investigate differences in values of operational measures to offer an even richer understanding of institutional variation in the net metering policy domain.

Drawing on the points above, we conclude this discussion by commenting on the generalizability of our findings to other domains of environmental sustainability, the value of the conceptual and methodological approaches we use for institutional analysis, and the relevance of engaging in assessments of institutional design in relation to environmental sustainability. Net metering policies are not unique kinds of environmental policies insofar as they contain multiple embedded elements that are configured in support of policy aims. Furthermore, net metering policies are among the clean energy-oriented policies that are gaining prominence for their promise to curb greenhouse gas emissions associated with energy generation, most of which are designed to encourage some kind of socio-technical transition (Geels 2010). By taking both factors together, our study applies a generalizable analytical approach to evaluating policy design. Specifically, we focus on common policy elements that operationalize variably across particular policy cases. This approach is applied to an increasingly popular kind of policy that is likely to remain salient in the context of global climate change. The conceptual and methodological approaches we use in our research, in addition to being readily transferable to other studies, reinforce the importance of understanding features of institutional design as part of any institutional analysis; that is, it underscores the importance of rigorous description of institutions as a basis for understanding institutional antecedents and/or outcomes. Finally, we suggest that because institutions are, and will continue to be, a primary mechanism through which governments seek to pursue environmental sustainability goals, applications such as ours that showcase how to engage in the institutional analysis of important sustainability-related concepts such as institutional diversity can offer important platforms for future research.

Conclusion

This edited volume is dedicated to the topic of institutional diversity and environmental sustainability. The pairing of these two concepts in the foci of the volume suggests that institutional diversity in environmental governance matters, arguably for many pragmatic reasons. Environmental issues are contextually sensitive; social and environmental factors differ across contexts, which has implications for how environmental issues manifest and evolve. Thus, there are no institutional panaceas that effectively transcend across contexts (Ostrom et al. 2007). Relatedly, institutional diversity is a logical byproduct of governments' choice to decentralize environmental policymaking. In federal systems like the US, this decentralization is manifest in the devolution of policymaking and implementation responsibility pertaining to a variety of environmental issues at the state level. Our study assesses the implications of decentralization on institutional design by investigating net metering policy in the United States, ultimately, toward a more expansive characterization of institutional diversity using the lens of robustness. The concept of robustness gives us multiple indicators to engage in intra- and inter-policy comparisons. We can assess how individual institutions (i.e., single policies) fare on robustness indicators (diversity, modularity, and redundancy), and we can also assess how different institutions (i.e., multiple policies) fare across the same. In linking diversity with robustness, we hope to stimulate additional research that seeks to explore conceptual and operational variants of institutional diversity toward more comprehensive assessments of the strategies through which governments are pursuing environmental sustainability.

References

Anderies, J.M. and M.A. Janssen. 2013. Robustness of social-ecological systems: implications for public Policy. Pol. Stud. J. 41: 513–536.

Arnold, G. 2015. When Cooperative Federalism Isn't: How U.S. Federal interagency contradictions impede effective wetland management. Publius J. Federalism 45: 244–269.

Basurto, X. 2005. How locally designed access and use controls can prevent the tragedy of the commons in a mexican small-scale fishing community. Soc. Natur. Resour. 18: 643–659.

Blomquist, W., T. Heikkila and E. Schlager. 2001. Institutional and conjunctive water management among three Western States. Nat. Resour. J. 41: 653–683.

Brady, U. 2020. Robust Conservation Anarchy: Comparing Treaty Institutional Design for Evidence of Ostrom's Design Principles, Fit, and Polycentricity. Ph.D. Thesis, Arizona State University, Tempe, AZ. ProQuest Dissertations & Theses Global.

Capano, G. and J.J. Woo. 2017. Resilience and robustness in policy design: a critical appraisal. Policy Sci. 50: 399–426.

Crawford, S.E.S. and E. Ostrom. 1995. A grammar of institutions. Am. Polit. Sci. Rev. 89: 582–600.

Engel, K.H. 2006. Harnessing the benefits of dynamic federalism in environmental law. Emory Law J. 56: 159–188.

Folke, C., S.R. Carpenter, B. Walker, M. Scheffer, T. Chapin and J. Rockström. 2010. Resilience thinking: integrating resilience, adaptability, and transformability. Ecol. Soc. 15: 20.

Frantz, C.K. and S. Siddiki. 2021. Institutional Grammar 2.0: A specification for encoding and analyzing institutional design. Public Admin. 99: 222–247.

Geels, F.W. 2010. Ontologies, Socio-Technical Transitions (to Sustainability), and the Multi-Level Perspective. Res. Policy 39: 495–510.

Ghorbani, A. and G. Bravo. 2016. Managing the Commons: A Simple Model of the Emergence of Institutions through Collective Action. Int. J. Commons. 10: 200–219.

Karapin, R. 2016. Political Opportunities for Climate Policy: California, New York, and the Federal Government. Cambridge University Press, Cambridge, UK.

Kharrazi, Ali, Yadong Yu, Arun Jacob, Nemi Vora and Brian D. Fath. 2020. Redundancy, diversity, and modularity in network resilience: applications for international trade and implications for public policy. Current Research in Environmental Sustainability 2: 100006.

Ostrom, E. 1990. Governing the Commons: The Evolution of Institutions for Collective Action. Cambridge University Press, Cambridge, UK.

Ostrom, E. 2005. Understanding Institutional Diversity. Princeton University Press, Princeton, NJ.

Ostrom, E., M.A. Janssen and J.M. Anderies. 2007. Going Beyond Panaceas. P Natl. Acad. Sci. USA 104: 15176–15178.

Schneider, A. and H. Ingram. 1988. Systematically Pinching Ideas: A Comparative Approach to Policy Design. J. Public Policy 8: 61–80.

Schneider, A. and H. Ingram. 1997. Policy Design for Democracy. University of Kansas Press, Lawrence, KS.

Siddiki, S., T. Heikkila, C.M. Weible, R. Pacheco-Vega, D. Carter, C. Curley et al. 2019. Institutional Analysis with the Institutional Grammar. Policy Studies Journal. DOI: https://doi.org/10.1111/psj.12361.

State Net Metering Policies Updated November 20, 2017. https://www.ncsl.org/energy/state-net-metering-policies

8

Necessary but Not Sufficient

An Examination of Two Areas of Multilateral Environmental Institutions

Archi Rastogi,[1,][*] *Jyotsna Puri*[2] *and Mark Hopkins*[3]

Introduction

One key reason for creating institutions to guide multilateral cooperation is to deal with well-recognised "global goods" challenges (Kaul and Blondin 2016). By definition, global goods are those that have implications for the entire world, and because of the "tragedy of the commons" (Hardin 1968; Coase 1960; Schelling 1978), there is little incentive for individual agents—including nation states—to work or take unilateral action for the generation, preservation, or mitigation of these goods. Common examples of such public goods are health, conflict, environment, trade, oceans, biodiversity, and data protection. Consequently, agents or sovereign nation-states have cooperated to build and sustain multilateral institutions that are mandated through common agreements to address these public good challenges and in what Schelling (1978) calls a Pareto-optimal solution or solutions that are to the benefit of all the actors. While multilateralism itself has several forms, this chapter is concerned with the definition used by Keohane (1990) as "the practice of coordinating national policies in groups of three or more states, through ad hoc arrangements or by means of institutions."

[1] Lee Kwan School of Public Policy, National University of Singapore and Independent Evaluation, Unit, Green Climate Fund175 Art Center daero, Songdo, Korea.

[2] International Fund for Agricultural Development, and Columbia University, Via Paolo di Dono 44, Roma, Italy.
  Email: j.puri@ifad.org

[3] Global Green Growth Institute21-15 Jeongdong-gil, Seoul, Korea.
  Email: markhopkins1994@gmail.com

[*] Corresponding author: archirastogi@gmail.com

This paper does not reflect the official views of the different institutions where authors are currently or previously employed. All errors are the authors' own.

In this chapter, we use a case study approach to illustrate the development of international environmental and climate institutions and examine their creation and evolution in the context of an increasingly complex world. We propose the use of complexity as a useful framework. Complex systems are defined as those that are characterised by heterogeneity (different types of agents/components), networks (or interconnectedness), and emergence (nonlinear interactions that lead to a system that is different from a simple and linear sum of its parts) (Holland 2014). We identify these characteristics in the global context and propose that complexity offers a useful normative framework to assess if current multilateral environment institutions are suitably designed. We maintain that it is clearly important for institutions also to demonstrate these features if they are located within a complex system and then examine whether current institutions are meeting this challenge. We use the example of institutions designed to deal with two global challenges: food and climate. We draw particularly from the more dynamic and contemporary case of two climate finance institutions: the Global Environment Facility (GEF) and the Green Climate Fund (GCF).

The chapter has the following structure: a brief review of the literature on institutions is followed by a historical overview of the development of international institutions from the post-World War II period onwards. We then examine the key multilateral institutions for food and agriculture, as well as the largest environment and climate finance institutions (the GEF and the GCF), their mandates, and their evolution in the context of the increasing complexity of our global system. Section IV discusses our findings in a more strategic context, and Section V concludes.[1]

Theories of Institutional Evolution

Theories of causes of institutional change and evolution have generally focused on two types of explanations: sudden, exogenously driven change and more gradual endogenous change (Gerschewski 2021).

Exogenously Driven Institutional Change

One branch of the scholarship considers institutional change as an exogenously driven process, such as one driven by the community or the state, or where constituent members of society may engage in collective action to change formal or informal rules. For instance, for Libecap (1989), the sources of rules for property rights are

---

[1] Many multilateral institutions are co-terminus with a specific organization. For instance, the GCF and GEF represent specific institutions as the operating entities of the financial mechanism of the UNFCCC. In this way, they cover a specific set of practices related to the transfer of resources across countries for specific mandates, primarily through programmatic tasks. They provide arenas for decision-making among representatives of sovereign governments, and possess characteristics such as formal rules and informal rules. This is consistent with the definition provided by North as "the rules of the game in a society, or, more formally, are the humanly devised constraints that shape human interaction" (North 1990). Yet, they also share features of organizations, in that they have executives and staff that come together for specific purposes. This chapter is concerned with the institutional dimensions: the rules and practices governing the behavior of nation-states. An examination of the organizational dimensions would require a closer focus on dimensions of effectiveness or efficiency, which is not the focus herein. Therefore, this chapter considers the dimensions that are institutional, rather than organizational.

rule-changing activities like contracting. Libecap (1989) suggests that institutional change is driven by a shift in exogenous parameters, which are driven by political rules, together with activities and perceptions of actors. Ostrom (2005) considers institutional change within a framework of operational rules, collective–choice rules, and constitutional rules; institutional change occurs when each actor assesses their costs and benefits from an institutional change, organise a minimum coalition to spur an overall institutional change.

Endogenously Driven Institutional Change

In another set of theories, institutional change is spurred not by the decisions of a central mechanism or a coordinated shift in rules. Instead, institutional change comes about due to uncoordinated choices of many individuals, and not necessarily collective choice or political processes (Coccia 2018). Much of this scholarship likens change in institutions to Darwinian evolution, where there is a natural selection of institutions both through the "selection of individuals endowed with the fittest temperament," and through the "adaptation of individual temperament and habits to the changing environment through the formation of new institutions" (Veblen 1899). Using the evolutionary framework, Young (1996) makes the argument that historical accidents could lead to the selection of particular types of institutions, and in the long term, the process will follow a "punctuated equilibrium"—a process marked by long periods of stability with rapid transitions from one established convention to another.

The latter branch has also been related to "new institutionalism" (DiMaggio and Powell 1983; Immergut 1998), where attention is paid to the role of external environments in shaping organisational change. DiMaggio and Powell introduce the concept of "institutional isomorphism" to help explain similarity among organisations in a specific organisational field (facing the same environmental conditions) and the way in which organisations become similar to their competitors (DiMaggio and Powell 1983). This approach rejects the older idea that institutions are a type of "rational-actor" and treats them instead as independent entities, placing them in a context of cognitive and cultural paradigms while also treating them as "supraindividual units of analysis" (or what we call networks and emergence) that cannot be "reduced to aggregations or direct consequences of individuals' attributes or motives." (DiMaggio and Powell 1983).

Change in Multilateral Institutions

Scholarship on international or multilateral institutions has focused less extensively on change itself. While studies of multilateral institutions and international organisations are longstanding (see Barnett and Finnemore 1999), analyses of institutional *change* are more recent. Oran Young's work has been particularly influential for studying institutional change in the context of what the author calls international regimes (particularly environmental regimes), which are equivalents of institutions in this chapter. Young (1982) defines a regime as a social institution that "governs the actions of those interested in specifiable activities … Regimes are social structures and should not be confused with functions." Importantly, international regimes involve international actors. Overall, he defines international regimes as sets of roles, rules, and relationships "that focus on specific problems and do not require centralized political organizations to administer them" (Young 1999). Young (1999) distinguishes between four types of institutional regimes based on the functions or tasks that they perform. These include regulatory, procedural, programmatic, and generative tasks (or a combination of these). For ease of illustration and comparison, all institutions referred to in this chapter perform programmatic functions primarily.

Altogether, these theories examine the creation and evolution of institutions that themselves have attributes of internal complexity. This scholarship explains several patterns, such as institutional change and isomorphism, effectiveness, fragmentation, and legitimacy (Young 2010; Biermann et al. 2020; Zürn 2018). We focus on patterns of institutional isomorphism because it helps us demonstrate the gaps among institutions that are unable to innovate and respond to emerging challenges.

In this chapter, we explore the cases of two sets of multilateral institutions related to food and climate, respectively. The institutions examined in this paper (the Rome-based UN food agencies, the GEF, and the GCF) primarily perform programmatic tasks combined with some normative work. These programmatic tasks involve pooling resources for the purpose of undertaking projects that "for one reason or another cannot be carried out on a unilateral basis" (Young 1999). At first, we explore the historical development of two multilateral institutions and then examine two propositions in how institutions respond to exogenous and endogenous change: one concerned with establishing new institutions, and the second related to the tendency for competition among institutions. These are responses to exogenous and endogenous factors, respectively, and we propose that, paradoxically, competition leads to a tendency that, despite being provided differentiated mandates at their inception, multilateral institutions have tended to converge in their structures, goals, and functions. We then propose using complexity theory as an organising framework for analysing the emergence of multilateral institutions, especially in the climate change space. Complexity theory offers a fruitful approach to this topic, as it highlights interconnectivity in messy, complex systems and the importance of variation and diversity for creativity, change, evolution, and emergence (Boulton 2010). We propose that complexity science provides a useful normative framework to explore whether international regimes are designed to tackle issues such as climate change can be said to constitute a "complex system" or whether institutions tend to resemble one another and are intrinsically unable to demonstrate innovation or ability to take risks. In particular, we provide insights from the development of multilateral institutions in the climate space and propose a normative direction for multilateral institutions.

The primary contribution of this chapter is to provide insights from modern multilateral institutions and practical implications. In doing so, the chapter makes use of some prevalent frameworks for the study of institutions. Ultimately, it also provides a suggested normative framework of complexity for policymakers to help design institutions; it also provides normative suggestions for the institutions themselves.

Multilateral Institutions Evolve

Global international institutions are created and changed in response to emerging needs and persistent challenges recognised by sovereign nations while also underscoring the challenge of global public goods—i.e., the fact that no single country has either the incentive to develop a response or an action, and, that it is in the interest of all countries to cooperate, but for single countries not to do so (Schelling 1978). In this section, we provide a brief historical overview of the development of institutions in response to challenges of peace, environment, and climate change.

While the history of cooperation among nation-states is extensive, modern multilateral institutions were created post-World War II. In July 1944, the Bretton Woods Conference resulted in agreements that, after ratification, led to the establishment of the International Monetary Fund (the IMF) and the International Bank for Reconstruction and Development (later part of the World Bank group), together laying the foundation for the international financial system and monetary relations in the post-war period. Almost in parallel, diplomatic conferences in 1944–45 led to the Declaration of the United Nations and the formal establishment of the United Nations (UN) in October 1945. These two events built upon previous developments and laid the foundation for modern multilateral systems. Multilateral institutions have proliferated since. Indeed, the UN system grew to encompass extant multilateral agencies, such as the International Labour Organization and the International Postal System and developed its own specialised agencies, such as the United Nations Development Program (UNDP) and the United Nations Environment Programme (UNEP). From aviation to vaccination and from nuclear armament to migratory species, there is a vast number of multilateral institutions today, all of which draw upon, to various degrees, the UN system. For instance, various UN agencies focus on gender and women. Three Rome-based agencies focus on various dimensions of food and agriculture—the Food and Agriculture Organisation of the UN (FAO), the International Fund for Agricultural Development (IFAD), and the World Food Programme (WFP). Beyond their themes, the agencies are also distinguishable by their orientation; for instance, the WFP focuses primarily on humanitarian aid along with the UNHCR, the UN refugee agency. UNICEF, also known previously as the United Nations Children's Fund, is a United Nations agency responsible for providing humanitarian and developmental aid to children worldwide. Many of these agencies have established means for further coordination in areas such as food, gender, climate, and health. The humanitarian area has several institutions that inform the overall cluster system. In the next sub-section, we discuss the origins of these institutions, organised by their mandates.

Post-War Challenges

In the aftermath of World War II, Allied governments established some major multilateral institutions that continue to provide a foundation for contemporary world governance. The Bretton Woods institutions and the UN were designed to address emerging challenges related to peacekeeping and economic cooperation, as well as to assist in post-war reconstruction and development. The preamble to the UN Charter, for instance, begins with a declaration "to save succeeding generations from the scourge of war," indicating the motivations and exigencies underpinning the establishment of these organisations (United Nations 1945). Since the founding of the Bretton Woods institutions and the UN, many globally relevant institutions have continued to be founded in response to shifting international challenges and priorities. While peacekeeping and security continue to be important priorities for the UN, the emergence of new issues has spurred the creation of more issue-specific institutions. The proliferation of various international regimes or governing arrangements constructed by states to deal with specific issue areas (including trade, monetary relations, and the oceans) has necessitated the development of institutions to monitor and operationalise such regimes (Frederking and Diehl 2015).

Food Agencies

One of the key areas that were recognised as urgent and important in the post-WWII period was hunger and poverty. Global hunger and increasing food insecurity and a recognition that these had global good attributes and cross-border externalities (food insecurity leads to migration, disease and nutrition burdens, and a lack of human capacity, mortality, and limited trading ability) led to the establishment of the Food and Agriculture Organisation (FAO) in 1945. Two additional organisations focusing on hunger and poverty issues emerged in the following decades. The World Food Programme (WFP), initially designed as an experiment to deliver food aid through the UN system, was established in 1961 and subsequently enshrined as a fully-fledged UN programme in 1965. The International Fund for Agricultural Development (IFAD) was established in 1977 following the 1974 World Food Conference in Rome and in response to the oil price crisis. Today FAO, the WFP, and IFAD take on different aspects of food-related assistance: FAO has traditionally taken on the role of generating and providing food-related technical leadership and developing normative standards; IFAD for delivering and ensuring food security while building markets in rural areas; and WFP for providing assistance in humanitarian contexts.[2] In practice, these goals often overlap, and frequently there is a change in roles as well as a push to develop intra-organisational capacity in areas that "belong" to the other organisation, mainly to deal with the different organisational cultures but also to serve the overall remit of the organisations themselves.

Environment Organisations

Multilateral institutions have also evolved in response to growing concerns about environmental damage. In 1968–69, the UN General Assembly adopted two resolutions that convened the United Nations Conference on the Human Environment

---

[2] See the respective website for each organization: https://www.fao.org/home/en; https://www.ifad.org/en/; https://www.wfp.org/.

in Stockholm in 1972. Twenty years later, the United Nations Conference on Environment and Development (UNCED) convened in Rio de Janeiro in 1992. These resulted in the Stockholm and Rio Declarations, respectively. The 1972 United Nations Conference on the Environment in Stockholm led to the first stocktake on the impact of human action on the environment globally (Handl 2012). The multilateral process in Stockholm underscored the pressing need for multilateral action on the environment and led to the creation of the United Nations Environment Programme (now called UN Environment). The Stockholm Declaration was the first to articulate a common approach to addressing the emerging challenges related to the environment, resulting in broad goals for countries and the globe (Handl 2012).

In the two decades following Stockholm, the global focus and activism on environmental issues have increased considerably. As the environmental imperative has matured and gained complexity, the UN General Assembly in 1987 adopted the "Environmental Perspective to the Year 2000 and Beyond" (General Assembly resolution 42/186, Annex) that provided "a broad framework to guide national action and international cooperation [in respect of] environmentally sound development" (Handl 2012). Two years later, in 1989, the General Assembly decided to convene the UNCED. The subsequent Rio Conference in 1992 provided more detailed "normative expectations regarding the environment..." and posited "the legal and political underpinnings of sustainable development" (Handl 2012).

Among the subjects of the Rio Conventions, climate change as a mandate has undergone more change and proliferation relative to the others. In 1992, the United Nations Framework Convention on Climate Change (UNFCCC) was negotiated as one of the three environmental treaties. With time, both within and outside of the UNFCCC, a number of climate-related treaties have come into effect, including the Kyoto Protocol, the Paris Agreement, the Warsaw International Mechanism for Loss and Damage associated with Climate Change Impacts (Loss and Damage Mechanism), and the Technology Mechanism.

These climate change treaties have, in turn, led to the development of new multilateral institutions to assist in their operationalisation. The GEF was established in 1991 and serves as a financial mechanism for the UNFCCC in addition to four other conventions. The Adaptation Fund was set up under the Kyoto Protocol in 2001, although it was not launched until 2007. The Climate Investment Funds (CIF)—comprised of the Clean Technology Fund and the Strategic Climate Fund—were established in 2008 at the request of the G8 and G20. During the 2010 UN Climate Change Conference (COP-16) in Cancun, the GCF was formally established to serve as an operating entity of the financial mechanism of the UNFCCC.

New Mandates and Convergence

In this section, we focus primarily on food agencies and environmental agencies to illustrate the ideas of mandate development and isomorphism. We present two propositions in this context.

Proposition 1: When new mandates appear/are agreed upon, existing institutions are frequently found inadequate, and this provides a reason for creating new institutions.

When shifts in global challenges and priorities lead to new mandates, existing multilateral institutions are examined for their suitability in realising these mandates; frequently, this leads to the creation of new institutions.

In the case of food agencies, the WFP, for example, was a modest food programme of the FAO in 1962, even after the expansion of its mandate in 1965 (where it was to include emergency food aid so it could contribute to agricultural and rural development on a continuing basis) (Talbot 1991). A sharp rise in world food prices in the early 1970s led to fears that the world food system was spiralling out of control, and the 1974 World Food Conference consequently focused its efforts on the global supply and prices of food. New institutions established in the aftermath of the World Food Conference reflected such concerns: the World Food Council (WFC) and the FAO Committee on Food Security sought to ensure the stability of global food supplies (Maxwell 1996). Criticism of the FAO—and specifically that it had failed to avert the world food crisis of the early 1970s—was a primary factor in the creation of the WFP (Talbot 1991). The work of the World Food Council was nonetheless suspended in 1993, with FAO and the WFP assuming many of its responsibilities.

Environment agencies similarly illustrate this case. The GEF was established in 1991 in parallel to the ratification of the Rio Conventions. It started as a pilot programme at the World Bank before becoming the first operating entity for the financial mechanism of the UNFCCC when the latter came into force in 1992. In 1994, the GEF moved out of the World Bank system to become a separate institution (Graham and Serdaru 2020). The relationship between the UNFCCC and the Council of the GEF was agreed in a memorandum of understanding (MOU).[3] The MOU outlined the nature of the relationship between the two bodies. Accordingly, the Conference of the Parties (COP) of the UNFCCC provides regular guidance on policies, programmatic priorities, and eligibility criteria for funding.[4] The GEF has a similar relationship with the other Rio Conventions and other multilateral environmental treaties. Nonetheless, by the late 2000s, the GEF was increasingly seen to be insufficient for several reasons (Graham and Sedaru 2020). First was the issue of governance. Given the origins of the GEF and the fact that its constituency-based voting system and replenishment rules mirror those of the MDBs in major ways, developing countries criticised the extent to which it is dominated by wealthy states (Graham and Serdaru 2020). This debate extends back to the 1980s and 1990s when institutions such as the World Bank and IMF made financial support or aid to developing countries contingent upon policy reforms, commonly referred to as "structural adjustment programmes." Because these programmes were almost always based on the opinions of external experts, they came to be viewed as one-sided and paternalistic (Buiter 2007; Burnside and Dollar 2000). By the late 1990s, however, the development community was increasingly questioning the effectiveness of conditional foreign aid. Easterly et al. (2003) pushed back against the argument that an externally determined, "good" policy environment was necessary to leverage the

---

[3] UNFCCC decisions 12/CP.2 and 12/CP.3. See https://unfccc.int/topics/climate-finance/funds-entities-bodies/global-environment-facility.

[4] Pursuant to Article 11, paragraph 1 of the UNFCCC.

full benefits of foreign aid, and the World Bank, in its 1998 report "Assessing aid—what works, what does not and why," called for greater dialogue between recipient and donor countries to enhance the effectiveness of aid. Within this context, the concept of country ownership came to occupy a more central position within the development agenda. The Paris Declaration on Aid Effectiveness (2005) emphasised the importance of the ownership of aid by developing countries, a principle that was subsequently reaffirmed and broadened to include a wider range of social groups through the Accra Agenda for Action (2008) and the Busan Partnership for Effective Development Cooperation (2011). Participants in these dialogues, representing both donor and recipient communities, committed to working together to build development strategies and to use developing country-level systems where appropriate (Buiter 2007).

A second major issue that highlighted some of the GEF's limitations was the growing scale of the climate change threat throughout the 2000s and a related ambition among countries to channel larger amounts of finance to help address this issue. A frequently mentioned shortcoming of the GEF in this regard was its broad mandate: the GEF serves as a financial mechanism to five conventions (the Convention on Biological Diversity, the UNFCCC, the Stockholm Convention on Persistent Organic Pollutants, the UN Convention to Combat Desertification, and the Minamata Convention on Mercury). This makes it impossible to focus efforts on channelling large volumes of finance to address the climate change issue alone. Moreover, the framework for climate finance under the UNFCCC was fragmented, making it insufficient to meet the growing financing needs of developing countries (Yamineva and Kulovesi 2013).

With increasing recognition and growing concerns surrounding both developing country ownership and the scale of the climate threat during the late 1990s and 2000s, it became clear that existing institutions such as the GEF were not capable of addressing these global needs. Such a recognition led to the creation of the GCF at the sixteenth meeting of the COP (COP-16) in Cancun, Mexico, in 2010, with Parties agreeing to establish the GCF as an operating entity of the Financial Mechanism of the UNFCCC.[5] A primary motivation for the creation of the GCF was that additional goals needed to be carried out (e.g., country ownership and direct access), which were not possible within the existing GEF model.

As written in the Governing Instrument of the GCF, the Fund is governed by a Board and is accountable to and functions under the guidance of the COP to support projects, programmes, policies, and other activities in developing country Parties using thematic funding windows.[6] The GCF Board is composed of an equal number of members from developing and developed country Parties (including representatives of relevant UN regional groupings and representatives from small island developing states and least developed countries). The GCF Board is responsible for funding decisions and operates with a relatively greater degree of autonomy than the

---

[5] UNFCCC decision 1/CP.16.

[6] UNFCCC decision FCCC/CP/2011/9/Add.1, Decision 3/CP.17, Annex (*Governing Instrument for the Green Climate Fund*).

decision-making bodies of the GEF and the Adaptation Fund. Board decisions are taken by consensus, and the Rules of Procedure, which supplement the governing instrument, provided that the questions be put to a vote if consensus is not reached (Bowman and Minas 2019). The parity in Board membership between developed and developing countries provides for stronger representation and voting rights for developing country Parties. This helps to address some concerns surrounding insufficient country ownership in institutions such as the GEF and World Bank.

In addition, the GCF is specifically focused on channelling financial resources from developed to developing countries and catalysing public and private climate finance. In contrast to the GEF, which serves five conventions, the GCF's targeted mandate means that it is well positioned to play a key role in the goal of mobilising USD 100 billion per year to address climate change needs in developing countries. The GCF is also mandated to support developing countries in implementing the 2015 Paris Agreement, including Article 2.1, that strives to make finance flows consistent with a pathway towards low emissions and climate-resilient development (GCF 2020).

The establishment of the institutions identified above is a demonstration of the responses to exogenous factors of institutional change. Using the framework provided by Ostrom (2005), it can be argued that when actors (in this case, nation-states) assessed the costs and benefits of prevalent institutions and then organised a minimum coalition to spur an overall institutional change, leading to the establishment of newer institutions with more specialised mandates. However, the institutions themselves demonstrate an inability to respond to exogenous factors proactively. Next, let us consider how these newly established institutions respond to endogenous change.

Proposition 2: Competition among institutions means institutions are always keen to find their own comparative advantage but not be too disruptive.

As demonstrated by the competition between the three Rome-based institutions as well as newer climate finance institutions, there are continual pressures for organisations to find their own comparative advantages without straying too far from the established norm. The latter inclination comes from the internal political economy of institutions (most leadership has a term limit, so the incentives to take on risks are limited), which leads to institutional risk aversion and, consequently, to convergence and institutional isomorphism. In this way, we suggest that one response of multilateral institutions to endogenous factors is to seek established institutional forms and resort to isomorphism.

The GEF was perceived to have a limitation in that it operated only through an exclusive set of eighteen eligible "GEF Agencies", comprising large international organisations (and three national organisations in China, Brazil, and South Africa) to carry out projects. The perceived and actual lack of control by national actors on programming decisions of these agencies led to the emphasis on "direct access" in the establishment of the Adaptation Fund and later the GCF. In this way, the mandate has continued to evolve. However, the nature and operation of direct access has remained similar among both the Adaptation Fund and the GCF and have not veered far from the institutional setup established under the GEF. Both the Adaptation Fund and GCF have provided direct access to organisations under conditions similar to those of the GEF. Indeed, the GCF portfolio continues to be dominated by GEF Agencies, thereby defeating the purpose of direct access.

Similarly, the GCF was provided a specific mandate related to the private sector, which included a thrust not only on additional mobilisation but catalysing the private sector to meet demands of climate change mitigation and adaptation. However, according to a recent evaluation, the private sector portfolio of the GCF continues to resemble that of other multilateral institutions (IEU GCF 2021). While mandates such as direct access and the private sector provide opportunities to undertake programming in drastically different areas through fundamentally different designs, institutions such as the GCF have not demonstrated the alacrity or ability to undertake disruptive change.

The establishment of the GCF in 2010 and the channelling of a greater volume of climate finance has placed pressure on the GEF to articulate its own strategic priorities and comparative advantage in order to distinguish it from other financial mechanisms. In a 2018 report, the GEF's Independent Evaluation Office (IEO) found that the primary comparative advantage of the GEF was its mandate as the principal financial mechanism of a number of multilateral environmental agreements (MEAs) and conventions, providing it with broad thematic coverage of environmental issues. Indeed, a survey conducted as part of the evaluation indicated that ninety-five percent of respondents "agree that the GEF's comparative advantage stems from its broad coverage of environment issues rather than any one specific issue area" (IEO GEF 2018). The report also argued that several features distinguish the GEF from other multilateral climate funds, including the provision of significant grant financing; the focus on developing enabling environments, such as through policy, legal, and regulatory reform, to support broader climate investment; and the emphasis on piloting potentially scalable technologies and financial approaches (GEF IEO 2018).

The GCF is also not immune to the pressures of competition among institutions and the need to define an institutional comparative advantage. A key problem for the GCF is that taking on an ambitious mandate also embodies key tensions that result in struggles over how to operationalise many of the provisions set out in its Governing Instrument. There are questions, for instance, around how the GCF's strategic vision to promote a paradigm shift toward low-emission and climate-resilient development pathways should be achieved. Some have warned that if poorly defined, this vision of transformational change "risks imposing a vague set of undefined pressures on developing countries" (Winkler and Dubash 2016). Similarly, Bertilsson and Thörn (2021) contend that in the context of a managerial governance style in which "guidance" to achieve paradigm shifts is provided to countries, country ownership becomes less about the bottom-up development of projects and instead refers to an institutional capacity to handle a project.[7] The GCF thus embodies what Kalinowski (2020) sees as a compromise between various groups, including developing countries that want more country ownership, the profit-seeking private sector seeking more

---

[7] These tensions are partly a result of the innovative nature of the GCF as an institution, given that it was designed to overcome criticisms directed at older international organizations, such as the World Bank.

business opportunities, and the civil society sector pushing for more openness and transparency in decision-making and operations.

As seen by the cases above, there is a tendency among newly established institutions to compete and resemble each other. This tendency relates to the responses to endogenous change. While responding to external environments, these new institutions have demonstrated a tendency for "institutional isomorphism" (DiMaggio and Powell 1983), wherein both GEF and GCF have evolved to become similar in their business model and delivery.

Discussion

With the emergence of new goals and/or mandates, we see a tendency for convergence and isomorphism on one side and a constant need to compete and define a comparative advantage among international institutions on the other. Such tendencies of multilateral institutions provide examples and illustrate the responses of multilateral institutions to exogenous and endogenous changes, respectively.

We find, however, that there is a resulting paradox from the responses of multilateral institutions to endogenous and exogenous factors. On the one hand, responses to endogenous factors lead to isomorphism and redundancy, and on the other hand, the strategies to address exogenous factors keep institutions from taking risks. Such trends consequently lead to unmet needs and therefore enduring gaps in attempts to serve key global goods in the world today. Two particular cases[8] in which this is evident are worth examining.

1. Loss and Damage

Most climate organisation are now experiencing pressures to address other emerging global needs, one of which is loss and damage. There is already a tendency to assign the challenge of addressing loss and damage to the GCF. In the guidance provided by the COP to the GCF, the COP invited the Board of the GCF "to continue providing financial resources for activities relevant to averting, minimizing and addressing loss and damage in developing country Parties" to the extent consistent with the existing investment, results framework, funding windows, and structures of the GCF. Additionally, it facilitates efficient access in this regard and in this context, takes into account the strategic workstreams of the five-year rolling work plan of the Executive Committee of the Warsaw International Mechanism for Loss and Damage associated with Climate Change Impacts.[9] The GCF has, however, been less forthcoming in this area. A report by the Stockholm Environment Institute finds that "such a funding mechanism is highly politically infeasible in the immediate term due to the institutional, structural and political barriers imposed by the climate finance

---

[8] As a postscript, it is worthy of note that both cases have developed in November and December 2022, during the writing of this paper. On Loss and Damage, the UNFCCC commissioned the establishment of an institution, while a new Biodiversity Fund was established by the CBD and is to be placed under the GEF.

[9] UNFCCC decision FCCC/PA/CMA/2019/6/Add.1, decision 6/CMA.2.

architecture" (Shawoo et al. 2021). The project-based funding model of the GCF and the lack of distinction between loss and damage and adaptation have been key factors impeding the needed consensus for the GCF to provide actively such resources.

2. Biodiversity Fund

Similarly, there is a developing international thrust on the establishment of a fund for biodiversity. For reasons similar to the ones on climate, there are limitations to the extent that the GEF and the Adaptation Fund can pursue this. These limitations relate mainly to the nature (country ownership) and the volume of resources available. Furthermore, there remains a perception among the five conventions served by the GEF that the mandate for this is rather weak.[10] Therefore, Parties to the CBD are exploring institutions such as the GCF to possibly circumvent the need to negotiate and establish a completely new institution. This would be expedient from the perspective of negotiations and would also mean that a biodiversity fund would be operational far sooner than in the case that a new institution needed to be established. However, for the time being, the common perception is that available institutions, both due to their current capacities and their ambitions, as well as the perception among staff that they would be stretched too thin, means that the challenge of accommodating this goal using extant institutions is seen as insurmountable.

The cases of loss and damage and the biodiversity fund provide examples of where multilateral institutions can be paralysed when faced with emerging mandates. Focused on endogenous factors and the related tendency for isomorphism, multilateral institutions in our example are not able to preemptively take into account emerging exogenous factors. This discussion is consistent with more recent critiques of multilateral institutions and public institutions, such as that offered by Mazzucato (2013), who argues that institutions are assessed as being averse to taking on critical roles with respect to the economy as a whole and the global system at large. According to Mazzucato, public institutions (of which multilateral institutions are a subset) must embrace their role as risk-takers and innovators—a role that has slowly and surely been increasingly forsaken in the post-WWII years. Her work focuses on the potential role that multilateral institutions can play in both coordination and market creation—a role that has been increasingly diminishing.

Complexity Theory and Multilateral Institutions

We organise the above understanding of multilateral institutions in a framework adopted from complexity science. Complexity science is a multidisciplinary scientific field that studies systems "with many interconnected parts" (Holland 2014). Systems with interconnected parts include both natural systems (such as ecosystems of varying sizes or multi-celled organisms) as well as social systems, including economic markets. Complex systems are characterised by a number of properties, such as that of "emergence" (wherein a system as a whole is greater than the sum of its parts), heterogeneity, networks, nonlinear behaviour (where the output

---

[10] As pointed out by the GEF IEO (2018).

of the system is not proportional to the input), and feedback loops (where a change in a variable causes either amplification or dampening of the initial change). Other features include "adaptive interaction", whereby agents in a system alter their behaviours based on the behaviours of other agents, and "hierarchical organisation", in which different phenomena occur at different levels of the hierarchy (DeCoste and Puri 2019). We argue that complexity science provides a useful framework for conceptualising and guiding the system of multilateral institutions, given its continually evolving nature and growing complexity. Specifically, three features of complexity science are important to consider in the context of modern multilateral institutions: heterogeneity and hierarchy, networks and interconnectedness, and emergence. These are briefly explored below.

Heterogeneity and Hierarchy

Complex systems, according to Wu (2013), tend to be structured in layers or levels—that is, they are hierarchically organised. Components at the same hierarchical level are likely to interact at faster rates than components located at different levels. Multilateral institutions examined in this chapter demonstrate that there is considerable heterogeneity and hierarchy. These institutions have expended considerable efforts and energy to distinguish themselves from one another and occupy relatively niche areas. For instance, climate institutions have different mandates on direct access and the private sector. The Rome-based agencies have established working groups to determine precise areas of work. However, such institutions have also demonstrated limitations in stepping far from the "equilibrium" that they currently occupy. The delineation and evolution of these institutions has been incremental, not yet leading to profound changes in the makeup of the overall system. In other words, while some heterogeneity is evident within the system of multilateral institutions, we have yet to see the emergence of a coherent whole[11]. In their responses to exogenous and endogenous factors, it might be useful to consider heterogeneity and hierarchy as desirable system-wide characteristics to guide the development of multilateral institutions.

Networks and Interconnectedness

Networks play an important role in better understanding complex systems. According to Caldarelli (2020), there are three main roles that network theory plays in complexity science: firstly, networks are a readily available "microscopic tool" for describing complexity; secondly, networks provide a tool to model the structure and dynamics of a vast number of phenomena; and thirdly, they offer a powerful visual tool. The second role highlighted by Caldarelli is of particular significance for the study of multilateral institutions. Network theory presents agents in the

---

[11] A coherent whole in the context of multilateral environment governance might be envisaged as a system in which institutions coordinate more extensively to cover gaps in existing work areas while also identifying emerging needs (e.g., loss and damage) and developing innovative approaches for addressing these needs.

system as nodes and the connections between them as networks, thus allowing for the analysis of behaviours and the frequency of interactions between nodes (DeCoste and Puri 2019). It helps observers understand, in other words, the nature of patterns, cooperation, and the spread of information within a networked system.

As discussed above, in response to new global challenges and the (perceived) inadequacies of existing institutions for addressing these challenges, they are thus linked in the sense that the experiences of existing institutions feed into the design of new ones. This has meant that new institutions tend to embody new mandates or areas of focus—for instance, on country ownership of development finance or to promote transformational change—that aim to distinguish them from existing institutions. In practice, however, issues of true interconnectedness and coherence within the broader system of multilateral institutions remain unresolved. Bowman and Minas have argued that in the case of the GCF, for example, greater interlinkages with other UNFCCC bodies—and specifically the Technology Mechanism—could help the GCF overcome challenges related to replenishment, governance, and oversight (Bowman and Minas 2019). Similarly, among the Rome-based agencies, although collaboration should seem like an obvious choice, this is mostly seen as a burden and an imposition rather than a practical solution to deal with manifold problems of global food insecurity (FAO et al. 2021). An emphasis on networks among multilateral institutions would allow for connectedness to be seen by policymakers as desirable and increase the effectiveness of institutions in responding to collective challenges. A recognition of connectedness can allow institutions to specialise in specific mandates while exercising synergies with comparators and competitors. For instance, to extend the example provided by Bowman and Minas (2019), it could be conceivable that a connected GCF would have the agility to respond to exogenous challenges created by fragmentation of replenishment and governance. Likewise, better-connected Rome-based agencies may be able to offer a coherent whole set of institutions to address food and agriculture-related challenges.

Emergence

Emergence is a defining property of complex systems and refers to a system whose whole is greater than the sum of its parts. The Earth's climate is an obvious example of such a complex system. Efforts to address climate change typically include multiple types of interventions given the scale of the problem and the need to achieve "transformational change".[12] However, having multiple interventions complicates the identification of causal links between specific interventions and their intended outcomes; these multiple interventions are indeed likely to produce "emergent" outcomes that would not necessarily occur if interventions were implemented individually (DeCoste and Puri 2019). In the case of multilateral institutions, the overall landscape of the system is yet to show the effect where the whole demonstrates properties not seen in individual institutions. While multiple climate finance institutions continue to operate with modest efforts for cooperation, the collective

---

[12] As is provided for in the GCF's mandate, for example.

interventions have yet to result in a transformation greater than the sum of individual interventions. Further, while challenges such as food, climate, and adaptation have continued to evolve rapidly, the institutions and their responses have not yet strayed far beyond the already-accepted body of norms to take on new properties, which might include, for example, demonstrating greater levels of innovation or risk.

So far, this chapter argued that while institutions respond to endogenous and exogenous factors, a tendency for "institutional imitation" defeats the initial purpose of new institutions. We have argued that complexity provides a useful normative framework whereby institutions can be expected to show attributes of heterogeneity, networks, and emergence. Currently, however, organisations that populate the complex environment and climate system are not showing these attributes and are instead imitating others who have entered the system first, showing classic symptoms of risk aversion. If, by design, multilateral institutions are expected to demonstrate heterogeneity, networks, and emergence, it may support these institutions to show associated attributes of risks and innovation. We used the example of the GCF to provide examples of this and concluded that there is an opportunity for institutions to evolve by taking risks and entering arenas not otherwise served and, in turn, doing less "imitation." Taking on the attributes of risk-taking and innovation is a global good itself and will make it far easier to arrive at a system tipping point and move us towards the paradigm shift that is critically required, especially in the global environment and climate space.

Conclusion

In this chapter, we first present a general overview of the development of international institutions from the post-World War II period onwards. We then examined the Rome-based agencies for food and agriculture and the environment and climate agencies (the GEF and the GCF) and sought to understand their evolution in the context of the increasing complexity of the global multilateral governance system. Gaps in this system were identified as areas that are a global public good but that are currently unserved by international institutions. Overall, we proposed that while multilateral institutions are focused on responding to endogenous factors, they may face a blind spot in exogenous factors. The use of complexity provides a useful framework that can allow multilateral institutions to avoid gaps, take risks, and demonstrate innovation. In conclusion, two normative directions are discernible for multilateral institutions:

1. Multilateral institutions must constantly pivot to remain relevant; otherwise, they will rapidly become obsolete.

It is clear that institutions must invest resources (time, learning, human resources) in their own development and growth and dynamically pivot their mandates, internal organisations, and capacities if they are to meet the needs of providing, supporting, and mitigating global public goods in the context of increasing complexity. This may involve "failing forward" and will involve innovation and experimentation in response to emerging needs. If the GEF were able to respond closely to the needs of the UNFCCC or the principles of country ownership, it is conceivable that there may not have been the thrust to establish a new institution for climate finance. Instead, it is possible that the mandate to mobilise USD 100 billion a year for climate change may have been accorded to the GEF, which would have yielded a far bigger mandate and profile.

Similarly, the GCF is possibly going to have to reexamine the suitability of its business model for the delivery of emerging mandates, such as loss and damage. This growth is compounded by the evident limitations of the business model of the GCF in delivering its core mandates, such as direct access, predictable finance, and a balance in its portfolio between adaption and mitigation investments.[13] Instead, the GCF has been able to channel finance only in the region of USD 10 billion every four years and already faces challenges in operations (IEU GCF 2019). If the GCF were indeed far larger, the achievement of diverse mandates simultaneously may have been possible. The GCF will have to continue to experiment and evolve if it is to retain its current mandates and not be rendered obsolete by an altogether new and potentially more relevant institution.[14] In other words, it will have to respond to exogenous factors and demonstrate some attributes of complexity, allowing it to take risks and remain relevant. The notion of constant pivot may be considered useful for the health of the individual institution, as well as the complex system of which the institution is part.

2. Diversity of institutions is not a bad thing, but we have witnessed risk aversion and limited innovation so far.

While the ever-evolving mandates of institutions and their diversity may exert internal pressures on these institutions, we are witnessing a paradox where, in response to exogenous factors, there is redundancy on the one hand and competition on the other (because of overlapping mandates). This means there are several global challenges that remain un-serviced even though there are a multitude of international institutions that could potentially address them. Barring a small proportion of the mandate or institutional design, a new generation of institutions tends to resemble the previous. This is the phenomenon of isomorphism. This allows institutions to learn from one another but means that there is limited innovation. Without innovation, institutions can become obsolete and irrelevant to an ever-changing landscape of development-related mandates.

---

[13] Let us also be clear that the mandate of the GCF is already complex, and one that has inherent tensions or trade-offs. For instance, if the GCF focuses primarily on direct access, it is possible that urgency and scale are less readily achievable. The Forward-looking Performance Review of the GCF counted no less than 18 different mandates within the GCF, many of which have inherent trade-offs (GCF IEU 2019). Many of these mandates were drawn up in the anticipation of an institution that channels finance in the order of hundreds of billion USD annually.

[14] Given the current limitations, however, the Forward-looking Performance Review (GCF IEU 2019) recommended the need to identify and strategize current priorities. Therefore, there are limits to the evolution of an institution such as the GCF.

References

Abbott, K., J. Green and R. Keohane. 2016. Organizational ecology and institutional change in global governance. International Organization 70(2): 247–277.

Barnett, M. and M. Finnemore. 1999. The politics, power, and pathologies of international organizations. International Organization 53(4): 699–732.

Bertilsson, J. and H. Thörn. 2021. Discourses on transformational change and paradigm shift in the Green Climate Fund: The divide over financialization and country ownership. Environmental Politics 30(3): 423–441.

Biermann, F., M. Van Driel, M. Vijge and T. Peek. 2020. Governance fragmentation. pp. 158–180. *In:* Biermann, F. and R. Kim (eds.). Architectures of Earth System Governance: Institutional Complexity and Structural Transformation. Cambridge University Press, Cambridge, UK.

Boulton, J. 2010. Complexity theory and implications for policy development. Emergence: Complexity and Organisation 12(2): 31–40.

Bowman, M. and S. Minas. 2019. Resilience through interlinkage: The Green Climate Fund and climate finance governance. Climate Policy 19(3): 342–353.

Burnside, C. and D. Dollar. 2000. Aid, policies, and growth. American Economic Review 90(4): 847–868.

Buiter, W. 2007. 'Country ownership': a term whose time has gone. Development in Practice 17(4-5): 647–652.

Caldarelli, G., S. Battiston, D. Garlaschelli and M. Catanzaro. 2004. Emergence of complexity in financial networks. pp. 399–423. *In:* Ben-Naim, E., H. Frauenfelder and Z. Toroczkai (eds.). Complex Networks. Springer, Berlin.

Caldareilli, G. 2020. A perspective on complexity and networks science. Journal of Physics: Complexity 1(021001).

Coase, R.H. 1960. The problem of social cost. Journal of Law and Economics 3(1): 1–44.

Coccia, M. 2018. An introduction to the theories of institutional change. Journal of Economics Library 5(4): 337–344.

DeCoste, S. and J. Puri. 2019. Complexity, climate change and evaluation. IEU Working Paper No. 02. Green Climate Fund. Songdo, South Korea.

DiMaggio, P. and W. Powell. 1991. Introduction. pp. 1–40. *In:* DiMaggio and Powell (eds.). The New Institutionalism in Organizational Analysis. University of Chicago Press.

DiMaggio, P. and W. Powell. 1983. The iron cage revisited: institutional isomorphism and collective rationality in organizational fields. American Sociological Review 48(2): 147–160.

Durlauf, S.N. 2012. Complexity. Economics, and public policy. Politics, Philosophy & Economics 11(1): 45–75.

Easterly, W., R. Levine and D. Roodman. 2003. New Data, New Doubts: A Comment on Burnside and Dollar's "Aid, Policies, and Growth" (2000). NBER Working Paper No. 9846.

FAO, IFAD. and WFP. (2021). Joint evaluation of collaboration among the United Nations Rome-Based Agencies. Rome. https://doi.org/10.4060/cb7289en.

Frederking, B. and P. Diehl. 2015. Introduction. pp. 1–12. *In:* Frederking and Diehl (eds.). The Politics of Global Governance: International Organizations in an Interdependent World, 5th edition. Lynne Rienner Publishers.

Gerschewski, J. 2021. Explanations of Institutional Change: Reflecting on a "Missing Diagonal". American Political Science Review 115(1): 218–233.

Gómez, E. and R. Atun. 2013. Emergence of multilateral proto-institutions in global health and new approaches to governance: analysis using path dependency and institutional theory. Globalization and Health 9(13): 1–17.

Graham, E. and A. Serdaru. 2020. Power, control, and the logic of substitution in institutional design: the case of international climate finance. International Organization 74(4): 671–706.

Green Climate Fund. 2020. Updated Strategic Plan for the Green Climate Fund 2020–2023. https://www.greenclimate.fund/document/updated-strategic-plan-green-climate-fund-2020-2023

Handl, G. 2012. Declaration of the United Nations Conference on the Human Environment (Stockholm Declaration), 1972 and the Rio Declaration on Environment and Development, 1992. Available at https://www.globalhealthrights.org/wp-content/uploads/2014/06/Stockholm-Declaration.pdf.

Hardin, Garrett. 1968. The tragedy of the commons. Science 162(3859): 1243–1248.

Holland, J.H. 2014. Complexity: a very short introduction. Oxford University Press. Oxford, UK.

Immergut, E. 1998. The theoretical core of the new institutionalism. Politics & Society 26(1): 5–34.

Independent Evaluation Office (IEO), Global Environment Facility. 2018. Evaluation of the GEF Partnership and Governance Structure. Available at: https://www.gefieo.org/sites/default/files/documents/reports/gef-partnership-governance-2017.pdf.

Independent Evaluation Unit. 2021. Independent evaluation of the Green Climate Fund's approach to the private sector. Evaluation Report No. 10, (September). Independent Evaluation Unit, Green Climate Fund. Songdo, South Korea.

Independent Evaluation Unit. 2019. Forward-looking Performance Review of the GCF. Independent Evaluation Unit, Green Climate Fund. Songdo, South Korea.

Kalinowski, T. 2020. Institutional Innovations and Their Challenges in the Green Climate Fund: Country Ownership, Civil Society Participation and Private Sector Engagement. Sustainability 12(8827): 1–13.

Kaul, I. and D. Blondin. 2016. Global public goods and the united nations. Chapter 2. *In*: Ocampo (ed.). Global Governance and Development. Oxford University Press. Oxford, UK.

Keohane, R.O. 1990. Multilateralism: an agenda for research. International Journal 45(4): 731–764.

Kingston, C. and G. Caballero. 2009. Comparing theories of institutional change. Journal of Institutional Economics 5(2): 151–180.

Krajnović, A. 2018. Institutional theory and isomorphism: limitations in multinational companies. Journal of Corporate Governance, Insurance, and Risk Management 5(1): 1–7.

Libecap, G.D. 1989. Contracting for Property Rights. Cambridge University Press. Cambridge, UK.

Maxwell, S. 1996. Food security: a post-modern perspective. Food Policy 21(2): 155–170.

Mazzucato, M. 2013. The Entrepreneurial State: Debunking Public vs. Private Sector Myths. Anthem Press.

North, D. 1990. Institutions, Institutional Change and Economic Performance (Political Economy of Institutions and Decisions). Cambridge University Press. Cambridge, UK.

Ostrom, E. 1990. Governing the Commons: The Evolution of Institutions for Collective Action. Cambridge University Press. Cambridge, UK.

Ostrom, E. 2005. Understanding Institutional Diversity. Princeton University Press. Princeton.

Schelling, T.C. 1978. Micromotives and Macrobehavior. W.W. Norton & Co. New York.

Shawoo, Z., A. Maltais, I. Bakhtaoui and S. Kartha. 2021. Designing a fair and feasible loss and damage finance mechanism. SEI brief. Stockholm Environment Institute. Stockholm, Sweden.

Talbot, R.B. 1991. The Four World Food Agencies in Rome as Political Institutions: Toward 2000. Transnational Law & Contemporary Problems 1(2): 341–392.

United Nations. 1945. United Nations Charter: full text. Available at https://www.un.org/en/about-us/un-charter/full-text.

United Nations. 1992. United Nations Framework Convention on Climate Change. Available at https://unfccc.int/resource/docs/convkp/conveng.pdf.

Watson-Grant, S., K. Xiong and J. Thomas. 2016. Country Ownership in International Development: Toward a Working Definition. Working paper. MEASURE Evaluation. https://www.measureevaluation.org/resources/publications/wp-16-164/at_download/document.

Veblen, T. 1899. The Theory of the Leisure Class: An Economic Study of Institutions. Macmillan. New York.

Winkler, H. and N.K. Dubash. 2016. Who determines transformational change in development and climate finance? Climate Policy 16(6): 783–791.

World Bank Group. 1998. Assessing aid—What works, what doesn't, and why. World Bank Policy Research Report. https://gsdrc.org/document-library/assessing-aid-what-works-what-doesnt-and-why/.

Wu, Jianguo. 2013. Hierarchy theory: an overview. pp. 281–301. *In*: Rozzi, R., S.T.A. Pickett, C. Palmer, J.J. Armesto and J.B. Callicott (eds.). Linking Ecology and Ethics for a Changing World: Values, Philosophy, and Action. Springer.

Yamineva, Y. and K. Kulovesi. 2013. The New Framework for Climate Finance Under the United Nations Framework Convention on Climate Change: A Breakthrough or an Empty Promise? pp. 191–223. *In:* Hollo, E., K. Kulovesi and M. Mehling (eds.). Climate Change and the Law. Springer.

Young, O. 1982. Regime Dynamics: The Rise and Fall of International Regimes. International Organization 36(2): 277–297.
Young, O. 1999. Governance in World Affairs. Cornell University Press.
Young, O. 2010. Institutional dynamics: Resilience, vulnerability and adaptation in environmental and resource regimes. Global Environmental Change 20: 378–385.
Young, O. 2021. Grand Challenges of Planetary Governance: Global Order in Turbulent Times. Edward Elgar Publishing.
Zürn, M. 2018. A Theory of Global Governance: Authority, Legitimacy, and Contestation. Oxford University Press.

9

Enabling and Bridging Institutional Diversity Through Polycentric Governance Structures to Advance Sustainable Development

The Case Study of the Arctic Council[a]

Jennifer Spence,[1], Rolf Rødven[2] and Nina Ågren[3]*

Introduction

The idea that a single global regime has the leverage or capacity to address complex, multi-scalar environmental issues, like climate change, biodiversity protection, pollution, and environmental contaminants, is easily refuted (Morrison et al. 2019; Ostrom 2010b). While ambitious commitments have been made through the United Nations Framework Convention on Climate Change (UNFCCC), global events confirm that action is needed by numerous institutions for the world to be able to address, mitigate and adapt to climate change. Ultimately, responding to the environmental challenges that face the globe requires the attention of a broad range

[a] This chapter uses data and focuses on the operations and activities of the Arctic Council up until early 2021. While it does not examine more recent developments, the authors believe the analysis still has relevance for understanding the Arctic Council and other institutions with polycentric governance structures.

[1] PhD, Director, Belfer Center Arctic Initiative, Harvard Kennedy School, 79 John F. Kennedy Street,Cambridge, MA USA 02138.

[2] Executive Secretary, Arctic Monitoring and Assessment Programme, The Fram Centre, Box 6606 Langnes, 9296 Tromsø, Norway.
Email: rolf.rodven@amap.no

[3] Executive Secretary, Emergency Prevention, Preparedness and Response Arctic Council Secretariat.

* Corresponding author: jennifer_spence@hks.harvard.edu

of experts and officials. By extension, it has been well argued that we need diverse institutions working at multiple levels (global, regional, national, sub-national and local) to effectively respond to these challenges (Goegele 2020; Lemos and Agrawal 2006; Ostrom 2010a, 2010b; Pattberg and Widerberg 2016). Instead, ecosystems of diverse institutions that are flexible and adaptable are recognised as critical for responding to the complex environmental issues that we face and supporting sustainable development.

Of course, the existence of diverse institutions simultaneously working at multiple levels to respond to the world's sustainable development challenges presents its own complications (van Zeben and Bobic 2019). Effectively mobilising, orchestrating and bridging the actions of these institutions to develop and implement complex, international policies is a critical governance challenge. This has inspired renewed interest by scholars to understand and consider the characteristics of polycentric governance systems and their relevance in responding to these complex issues. By extension, a discourse has emerged to consider under what conditions polycentric governance systems should be introduced or promoted and how best to apply polycentric governance approaches (Goegele 2020).

Beginning with a review of the attributes of polycentric governance, we know these systems are defined as having multiple centres of authority that function with some degree of autonomy but take each other into account through the coordination of processes and an overarching system of rules (and Tarko 2012; McGinnis and Ostrom 2011; Morrison et al. 2019; Ostrom 2010b; Pahl-Wostl and Knieper 2014). Polycentric governance systems can be understood as a web of horizontal and vertical relationships where authorities at different scales seek to strike a balance between centralised and decentralised organisation (Carlisle and Gruby 2019).

Building on this literature, to what extent can the concept of polycentric governance enhance our understanding of institutions that internalise diverse expertise, perspective and needs? Can the idea of polycentric governance structures provide insights on how to assess and potentially enhance institutional effectiveness? There is an emerging literature that examines these questions in relation to complex policy issues and international institutions. For example, van Zeben and Bobic (2019) adopted polycentric governance to study the European Union and Geogele (2020) tested the explanatory power of this concept to inform our understanding of the United Nations. In this newer literature, the discussion shifts from examining polycentric governance systems that join diverse institutions in a particular policy field to considering the institutional structures and systems that enable the accommodation of diverse expertise, perspectives and needs within an institution.

The Arctic Council's unique institutional configuration and governance features provide another empirical case to test further and advance our understanding of international institutions with polycentric governance structures. Similar to van Zeben and Bobic's (2019) study of the European Union and Geogele's (2020) analysis of the United Nations, we seek to test the explanatory power of the concept of polycentric governance to shed light on the internal workings of the Arctic Council. In particular, this case study lends itself to further exploring the applicability of the concept to understanding international institutions with diffused authority structures and how these institutions internalise polycentricity. This case study seeks specifically to consider the relationship between the form and function of the Council and how polycentricity enables it to engage in environmental protection and sustainable development work. However, perhaps the more important contribution of this chapter is to deepen our understanding of how diversity can be enabled and mobilised within an institution. We will examine the work and institutional features of the Arctic Council with a particular focus on how its governance structures bring together a broad array of experts and knowledge holders, bridge different types of expertise and knowledge systems, and orchestrate diverse expertise to inform and influence policy-making.

The Arctic Council: A Case Study of Polycentric Governance

Established in 1996, the Arctic Council is an intergovernmental forum with a mandate focused on supporting sustainable development and environmental protection in the Arctic. The Council's membership is made up of the eight states with jurisdictions in the Arctic (Canada, Kingdom of Denmark, Finland, Iceland, Norway, Russian Federation, Sweden, and the United States) and assigns special status to six Indigenous organisations as "Permanent Participants" (Aleut International Association, Arctic Athabaskan Council, Gwich'in Council International, Inuit Circumpolar Council, Russian Association of Indigenous Peoples of the North, and Saami Council). The Arctic Council also provides Observer status to a growing number of non-Arctic states, intergovernmental organisations and non-governmental organisations.

The Council is led by the Ministers of Foreign Affairs of the eight Arctic states, with the involvement of the Indigenous Permanent Participants. These Ministers meet every two years, when the chairship of the Council rotates between Arctic states, to approve the work that has been completed by the Council and its work plan for the next two years. Between these biennial ministerial meetings, responsibility for the leadership of the Council is delegated to Senior Arctic Officials, appointed diplomats from each country, and senior officials representing the Indigenous Permanent Participants, who are tasked with leading the work of the Council for the duration of each chairship (see Figure 1).

The Council is supported by six working groups: The Arctic Contaminants Action Programme (ACAP), the Arctic Monitoring and Assessment Programme (AMAP), the Conservation of Flora and Fauna Working Group (CAFF), the Emergency Prevention, Preparedness and Response Working Group (EPPR), the Protection of the Marine Environment Working Group (PAME) and the Sustainable Development Working Group (SDWG). Ministers may also establish task forces or expert groups to undertake fixed-term initiatives for the Council, as needed. Examples of fixed-term expert groups and task forces that have been established include black carbon and methane, telecommunications infrastructure, Arctic marine cooperation, and search and rescue.

With growing global attention on the Arctic because of both the prominent impacts of climate change (ACIA 2005; AMAP 2017d) and its economic potential as a new frontier of resource extraction and shipping (AMAP 2017a, b, c), the Arctic Council and its role in sustainable development and environmental protection in the Arctic has gained increasing global attention. However, the institutional design and

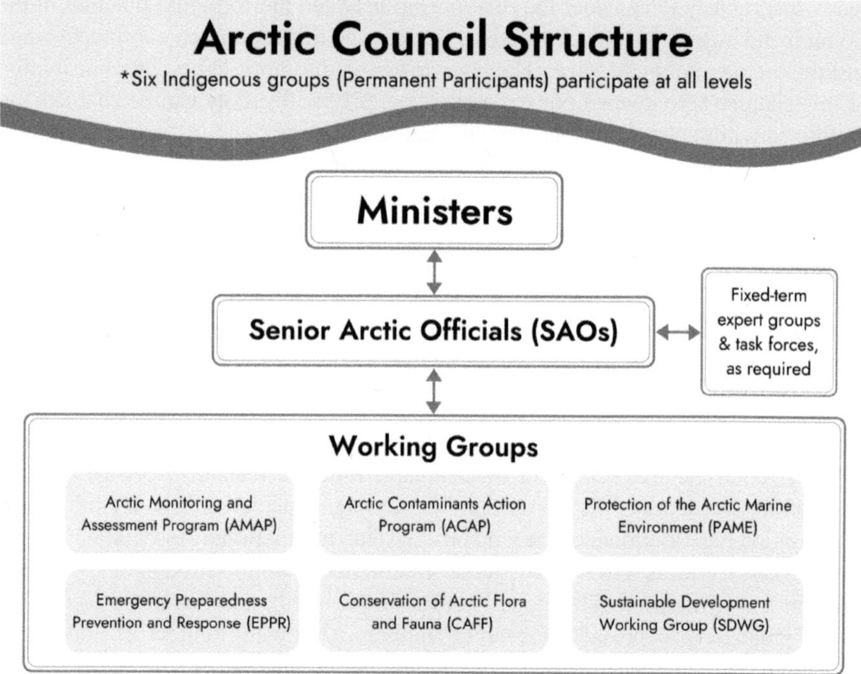

Figure 1: Organisational structure of the Arctic Council.

inner workings of the Arctic Council remain poorly understood by many researchers and even officials involved in the work of the Council.

Considering the attributes of polycentric governance, we propose to identify and assess the existence of key attributes of polycentric governance within the Arctic Council. In particular, we consider to what extent the Arctic Council's working groups represent "centres of authority" with "some degree of autonomy" to identify and deliver on their priorities. We also consider the role of Arctic Council Ministers, Senior Arctic Officials (SAOs), Indigenous Permanent Participants and the Arctic Council Secretariat to provide the "overarching system of rules and strategic direction."

To support this analysis, we begin by examining three of the Arctic Council's working groups—the Arctic Monitoring and Assessment Programme (AMAP), the Emergency Prevention, Preparedness and Response Working Group (EPPR) and the Sustainable Development Working Group (SDWG). As we will illustrate here, these three working groups differ significantly in their mandates and processes, reflecting institutional diversity within the Arctic Council. For each, we provide an overview of their mandate, how they are organised, and the scope of their work. To deepen the analysis, we also examine the areas of expertise of the delegates currently involved in each working group. To do this, we reviewed and coded the areas of expertise of the experts and knowledge holders currently involved in the three working groups by reviewing the biographies and curricula vitae of each delegate. Recognising that delegates involved in the working groups often hold expertise in multiple areas, each individual was coded for multiple areas of expertise, as appropriate.

In this chapter, we examine the range of work, the distinctive governance features and the diverse expertise that exists within the Arctic Council. We seek to demonstrate the Council's facility as a case study for polycentric governance within an international institution. We then consider the Council's governance mechanisms that serve to orchestrate the work of the Council and enable horizontal and vertical bridging of ideas and activities, recognising that this ability to draw the different authority centres of the Council together is critical to its institutional effectiveness.

Arctic Monitoring and Assessment Programme

The creation of the Arctic Monitoring and Assessment Programme (AMAP) precedes the establishment of the Arctic Council. AMAP was launched in 1991 as part of the Arctic Environmental Protection Strategy (AEPS) and was established as a pan-Arctic monitoring programme with a mandate "to monitor the levels of, and assess the effects of, anthropogenic pollutants in all components of the Arctic environment" (AMAP 2019). The AEPS and its working groups, including the AMAP, were subsumed into the Arctic Council at its creation in 1996. More recently, AMAP's mandate has evolved and expanded to include:

"...monitor[ing] and assess[ing] the status of the Arctic region with respect to pollution (e.g., persistent organic pollutants, heavy metals, radionuclides, and other chemicals of emerging Arctic concern) and climate change issues by documenting levels and trends, pathways and processes, and effects on ecosystems and humans, and by proposing actions to reduce associated threats for consideration by governments (Arctic Council 2018)."

AMAP is guided by its strategic framework (AMAP 2019), and strategic decisions are made by appointed Heads of Delegation from Arctic states, which primarily are officials from each state's Ministry of Environment or their Directorates and Indigenous Permanent Participants, and led by a board consisting of the Chair and three Vice-Chairs. Arctic Council observers are invited to participate in the AMAP's meetings and work, for instance, by nominating experts to join AMAP Expert Groups and non-Observers can contribute expertise, where appropriate (AMAP 2019). AMAP's work plan is developed by its Heads of Delegation based on policy relevant questions and experts presenting relevant knowledge gaps. Arctic Council observers are welcome to comment; however, decisions are made by Heads of Delegation of Arctic states with the involvement of Indigenous Permanent Participants.

AMAP's work is supported by a dedicated secretariat established in 1991, which was originally located in Oslo, Norway, and relocated to Tromsø, Norway, in 2018. AMAP's secretariat is well-established, and in 2021, it consisted of six secretariat staff members.

AMAP's main deliverables are thematic peer-reviewed scientific assessments. Since its first report on Arctic Pollution Issues in 1998 (AMAP 1998), AMAP has published 34 scientific assessment reports, on topics ranging from human health, climate change, ocean acidification, oil and gas, pollution, chemicals of emerging Arctic concern, adaptation actions as well as microplastics. The reports are authored by experts nominated to the AMAP expert groups, including expert groups on climate, Short Lived Climate Forcers (SLCF), Permanent Organic Pollutants

(POPs), mercury, ocean acidification, radioactivity and human health. The reports are subjected to a national data check by the AMAP Heads of Delegation to ensure that relevant data and policy-relevant questions are included, but the scientific reports are not subjected to national negotiations. AMAP reports are peer-reviewed by independent experts, similar to other scientific journals.

These assessments provide the scientific basis for recommendations on Arctic environmental issues that are addressed to the Arctic Council Ministers and Senior Arctic Officials. In addition, the assessment process is coordinated with other international processes, feeding data and information to international bodies such as the Intergovernmental Panel on Climate Change, UN Environment, Stockholm Convention, Minamata Convention and Long-range Transboundary Air Pollution (Platjouw et al. 2018; Rottem et al. 2020; Steindal et al. 2021). These comprehensive reports are also condensed into summaries for policymakers that include a scientific summary of key findings and recommendations for policymakers consideration.

The areas of expertise of delegates involved in AMAP work are outlined in Table 1. This analysis highlights that AMAP is primarily supported by natural scientists. The areas of expertise represented within AMAP can broadly be understood through two perspectives. Firstly, how trends and levels in pollution and climate change impact the Arctic environment, and secondly, how these environmental issues

Table 1: Areas of expertise of delegates involved in AMAP work.

Area of Expertise	Percentage of Delegates
Pollutants and Contaminants	45%
Earth and Ocean Sciences	29%
Arctic Studies	29%
Climate	26%
Environment	20%
Aquatic Ecology	17%
Atmospheric Science	17%
Health (human and animal)	13%
Toxicology	12%
Ecology	9%
Chemistry	9%
Modelling	7%
Biology	6%
Indigenous Studies	4%
Food Systems	3%
Energy	3%
Engineering and Technology	3%
Sustainable Development	2%
Emergency Management	1%

Note: Each delegate is coded using publicly available biographies and CV. They may contribute expertise in more than one area and have been coded accordingly.

impact the lives of those living in the Arctic. In particular, we observe that 45% of AMAP delegates hold expertise related to pollutants and contaminants, in which they often specialise in mercury and other persistent organic pollutants, radionuclides, and heavy metals. This is consistent with the topics addressed in AMAP's scientific reports. That being said, these experts often seek to understand how pollutants and contaminants interact with the living environment or with humans and animals. Due to this, earth and ocean sciences specialisations contribute to a sizeable portion (29%) of AMAP's expertise, with subcategories including, but not limited to, Arctic glaciology and snow and marine and freshwater systems. Expertise related to human interaction makes up a significantly smaller percentage of expertise present in AMAP and includes diverse disciplines, including Indigenous practices, health (which includes human, animal and epidemiology), toxicology, climate, and sustainable development, among others. However, this diversification may be confounded by the expert's educational background; while both a toxicologist and a medical doctor may both be working with human health, they may categorise themselves as a pollutant and health specialist, respectively.

Emergency Prevention, Preparedness and Response Working Group

The Emergency Prevention, Preparedness and Response Working Group (EPPR) was also established under the Arctic Environmental Protection Strategy in 1991. Originally, EPPR was established as an experts' forum with the main objectives of reviewing existing cooperation in the Arctic related to emergency preparedness and to consider and recommend necessary and adequate systems of cooperation (Arctic Environmental Protection Strategy 1991). EPPR's mandate, as outlined in the EPPR Strategic Plan (EPPR 2016), contributes to the prevention, preparedness and response to environmental and other emergencies, accidents, search and rescue, threats from unintentional releases of pollutants and radionuclides, and to consequences of natural disasters in the Arctic. EPPR is not an operational response organisation and focuses on developing guidance, best practices and risk assessment methodologies, exercise and training coordination and information sharing. The topics EPPR focuses on have shifted over time depending on the priorities and tasking of Arctic states, developing and expanding as the activities, size and capacity of the EPPR have grown.

The secretariat functions for EPPR initially rotated with the working group's chairship, as there was no permanent secretariat. Discussions were initiated in 2011 about the need for more consistent and robust secretarial support to the EPPR that could be provided as part of a new permanent Arctic Council Secretariat. In 2013, the permanent Arctic Council Secretariat, based in Tromsø, Norway, became operational and began providing secretarial support to EPPR.

The EPPR Strategic Plan sets the framework for the EPPR work, and a biennial work plan follows the Arctic Council Chairship cycle. The work plan is developed by the Heads of Delegation, appointed by the Arctic states and Indigenous Permanent Participants to represent their respective states and organisations in EPPR. Many EPPR Heads of Delegation work in relevant government ministries or agencies dealing with emergency response and preparedness. Identified research needs and knowledge gaps contribute to project planning. Project proposals are developed by the Arctic states or Indigenous Permanent Participants, either with one state or Permanent Participant leading the project or with two or more co-leads. EPPR projects must be approved by the consensus of EPPR's Heads of Delegation. Observers take part in EPPR work where appropriate and relevant for them.

EPPR maintains and supports the implementation of two of the three international agreements that have been negotiated under the auspices of the Arctic Council: the Agreement on Cooperation on Aeronautical and Maritime Search and Rescue in the Arctic (Arctic Council 2011) and the Agreement on Cooperation on Marine Oil Pollution Preparedness and Response in the Arctic (MOSPA) (Arctic Council 2013). EPPR is responsible for the technical and administrative management of the agreements, such as revising and updating the MOSPA Appendix IV Operational Guidelines, thus requiring the participation of appropriate experts by the Arctic states, who are the parties to the agreements. EPPR also coordinates exercises associated with these agreements, documenting after-action reports and analysing lessons learned. These activities contribute to building cooperation and enhancing response capabilities in the Arctic.

EPPR organises its work with a recognition that the Arctic environment is unique, that inhabitants are most affected by the risks and threats that emergencies pose in the region, and that due to long distances and harsh conditions, local communities may be the initial responders before a regional, national or multilateral response is possible. EPPR projects range from circumpolar wildfire preparedness and response to environmental, social and economic consequences of oil spills to small communities. Projects contributing to developing circumpolar risk assessments are part of the EPPR project portfolio, similar to projects contributing to research and development in EPPR's core area of emergency prevention, preparedness and response. Outreach and communication of risks and supporting the preparedness of Arctic inhabitants while contributing to emergency prevention are also important priorities.

EPPR currently has three expert groups that support its work. The Search and Rescue Expert Group (SAR EG) focuses on Arctic SAR-related projects and conducts activities related to the SAR Agreement, assesses its implementation and builds on experiences and research. The Marine Environmental Response Expert Group (MER EG) coordinates and conducts MOSPA-related activities, training, exercises and workshops. Projects and initiatives under the MER EG focus on oil spill prevention, preparedness and response. The third and newest Expert Group on Radiation (RAD EG) was established by EPPR in late 2019 and is mandated to contribute to and facilitate EPPR work on radiological and nuclear emergencies. Besides maintaining the agreements, expert groups in EPPR are a way to organise expertise and work the group undertakes, as guided by the Arctic states and Indigenous Permanent Participants.

In line with EPPR's three expert groups, Table 2 highlights that there are currently five key areas of expertise present in EPPR: emergency management (30%), security and defence (25%), maritime and marine studies (25%), environment (20%), and energy science (18%). Almost every delegate in EPPR has expertise in at least one of these key areas and often interacts with one of the other areas of expertise. The intersection of emergency management and environment, energy, or pollutants and

Table 2: Areas of expertise of delegates involved in the EPPR work.

Area of Expertise	Percentage of Delegates
Emergency Management	30%
Security and Defence	25%
Maritime and Marine Studies	25%
Environment	20%
Energy Science	18%
Arctic Studies	12%
Pollutants and Contaminants	12%
International Affairs	7%
Ecology	5%
Civil Affairs	4%
Law and Policy	3%
Indigenous Studies	3%
Earth and Ocean Science	2%
Sustainable Development	2%
Culture and Identity	1%
Management and Mobilisation	1 %
Economics	1%

Note: Each delegate is coded using publicly available biographies and CV. They may contribute expertise in more than one area and have been coded accordingly.

contaminants is most commonly found. This can be explained by the EPPR's focus on these types of incidents within the Arctic. A small percentage of other areas of expertise are also observed in the EPPR depending on the nature and topics taken on, and as the working group has continued to grow over time, it will be interesting to observe how the diversity of expertise may expand further with emerging Arctic issues.

Sustainable Development Working Group

The Sustainable Development Working Group (SDWG) was established with the creation of the Arctic Council in 1996; however, its terms of reference were not adopted until 1998, making it one of the "younger" working groups of the Arctic Council. In these terms of reference, the SDWG was mandated to:

"propose and adopt steps to be taken by the Arctic states to advance sustainable development in the Arctic, including opportunities to protect and enhance the environment, and the economies, cultures and health of Indigenous communities and of other inhabitants of the Arctic, as well as to improve the environmental, economic and social conditions of Arctic communities as a whole (Arctic Council 1998)."

This broad mandate has been more colloquially branded as the "human dimension" of the Arctic, which stands in contrast to the mandates of the other working groups that have historically been more focused on environmental protection. However, over time, all working groups have increasingly placed more attention on issues and needs that are important to Arctic peoples.

Since its inception, the SDWG has received secretariat support through a professional services contract with the Government of Canada for an Executive Secretary. In 2019, Canada committed additional funding to establish a permanent secretariat in Quebec, Canada, that will substantially increase the resources and capacity available to support the work of the SDWG. It is expected that this new secretariat will be launched in 2022.

SDWG has the closest relationship to the leadership of the Arctic Council. Initially, it was Senior Arctic Officials (SAOs) and political leaders from Indigenous Permanent Participants that made up the membership of the SDWG. Over time, more distance has been established between the SDWG and the Arctic Council leadership; however, a large majority of SDWG Heads of Delegation are still diplomats and political leaders. This makes SDWG different from the other working groups because the membership is not made up of experts themselves, rather they serve as a focal point to draw in relevant expertise to contribute to the SDWG's work.

SDWG has two expert groups—one focused on human health and the other on social, cultural and economic issues. Representatives on both these expert groups are appointed by the SDWG Heads of Delegation and are primarily academics from domestic institutions with relevant expertise and interests. These expert groups do some standalone research and work; however, their primary role has been to be to provide advice, feedback and support to the SDWG.

SDWG adopted a strategic framework to guide its work in 2017, and it produces work plans for each Arctic Council Chairship (every 2 years); however, it is primarily a project-driven working group. An Arctic state or Indigenous Permanent Participant can propose a project and with the co-leadership of two Arctic states and the endorsement by consensus of the SDWG Heads of Delegation, the project can proceed. Occasionally, Arctic Council Senior Arctic Officials will task the SDWG with particular work; however, it is largely driven by a bottom-up approach to setting priorities.

SDWG projects have covered a broad range of topics, including but not limited to renewable energy, gender, human health, Arctic economies, Indigenous languages, suicide prevention and mental health, food, Indigenous and local knowledge, waste management, economic assessments and environmental impact assessments. The interests and involvement of Indigenous Peoples are strongly prioritised in the SDWG's work with many projects covering topics proposed by Indigenous Permanent Participants and regular involvement of Permanent Participants in individual SDWG projects.

SDWG is primarily supported by social scientists, and its areas of expertise are diverse and intersectoral (Table 3), with 80% of its delegates representing two or more disciplines, and 48% of those representing an intersection of more than three disciplines. That being said, the most commonly shared areas of expertise in the SDWG are Arctic studies, international affairs, environment, and Indigenous relations, with anywhere from 25% to 45% of members being expert in each discipline. A number of experts within the SDWG specialise in human experiences in the Arctic

Table 3: Areas of expertise of delegates involved in the SDWG work.

Area of Expertise	Percentage of Delegates
Arctic Studies	44%
International Affairs	33%
Environment	26%
Indigenous Studies	26%
Sustainable Development	17%
Culture and Identity	12%
Climate	12%
Health (human, animal)	11%
Management and Mobilisation	10%
Economics	7%
Earth and Ocean Science	6%
Law and Policy	6%
Resource Management	6%
Emergency Management	3%
Engineering and Technology	3%
Gender Studies	3%
Security and Defence	3%
Energy Science	3%
Food Systems	2%
Biology	2%

Note: Each delegate is coded using publicly available biographies and CV. They may contribute expertise in more than one area and have been coded accordingly.

and study international and local socioeconomic or health related challenges. Due to this, experts most commonly specialise in either environment or climate, in addition to Indigenous affairs, culture and identity, health or economics.

Enabling Diversity Through the Arctic Council Working Groups

This brief introduction to the Arctic Council and the distinct characteristics of three of its working groups demonstrates that, while AMAP, EPPR and SDWG clearly advance the Arctic Council's mandate of sustainable development and environmental protection, they have unique mandates and decision-makers organise themselves differently to meet their goals and objectives, have varying levels and types of secretariat support, and respond to very different policy issues. This confirms that organisational diversity exists within the Arctic Council and confirms the working groups as distinct centres of authority that function with a certain degree of autonomy.

We also presented an analysis of areas of expertise represented in each working group, affirming the distinct communities of experts and knowledge holders that have formed around each of these authority centres. Before moving on to a consideration of the institutional mechanisms that connect and bridge these working groups, we want to further elucidate this diversity by comparing the areas of expertise within the three working groups.

Of the 28 areas of expertise identified, only eight general areas of expertise are represented in all three working groups (Arctic studies, environment, earth and ocean science, economics, emergency management, energy, Indigenous studies, and sustainable development). Of the remaining 20 areas of expertise, SDWG and AMAP share five common areas (biology, climate, engineering and technology, food systems, and health sciences), AMAP and EPPR share two common areas (ecology and pollutant and contaminants), and EPPR and SDWG share five common areas (culture and identity, international affairs, law and policy, management and mobilisation, security and defence).

To further illustrate the distinctive expertise and communities that exist within AMAP, EPPR and SDWG, Figure 2 uses a bubble graph to provide a comparison of the top five areas of expertise represented in each working group. In addition, the

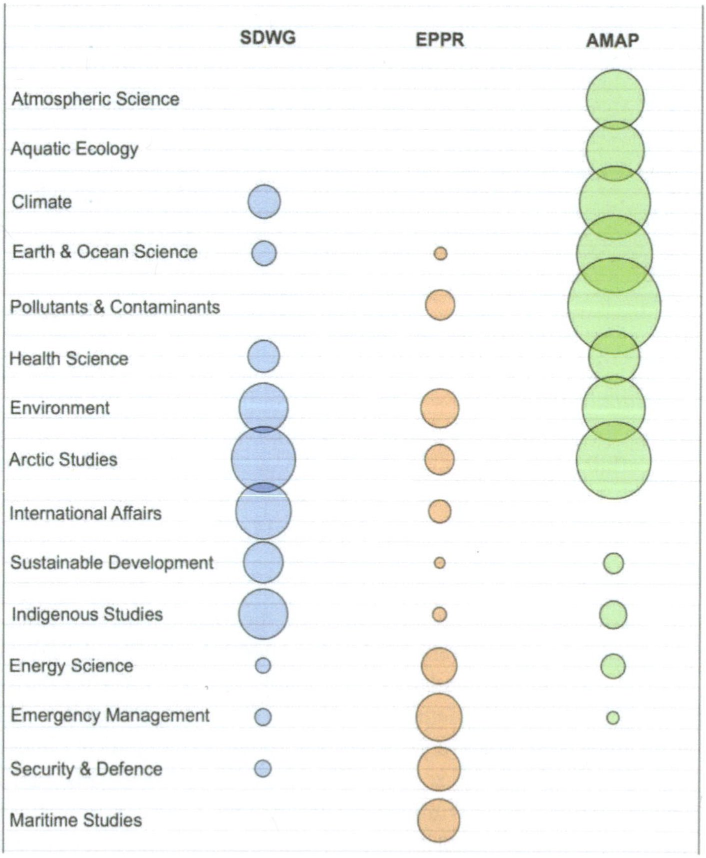

Figure 2: Areas of expertise by working group. Note: The top five areas of expertise for each working group are presented in this figure. The size of the bubble represents the number of experts within the working group with that area of expertise.

size of each bubble represents the number of delegates with those expertise. This figure highlights that there are some overlaps, but more evidence of distinct areas of expertise between these three working groups. It also demonstrates that AMAP engages the largest number of experts and knowledge holders relative to EPPR and SDWG.

This analysis confirms that the working groups depend on and build relationships with distinct expertise and communities of experts to advance their work. In this context, the Arctic Council's polycentric governance structure is a means to allow each working group, as an authority centre, to develop its own systems and structures for operating that align with the organisational norms and cultures of its experts and reinforce a sense of community. This ability to foster distinct communities within each working group acts as bonding systems between the Arctic Council and the networks of experts that it connects with and depends on to conduct its work (Hamilton and Lubell 2019). The flexibility and adaptability that is provided by a polycentric governance structure serves to position the Arctic Council as a focal point that brings together diverse communities of experts and knowledge holders to advance environmental protection and sustainable development and serves as a conduit between these communities and policymakers.

To recognise the Arctic Council as having a polycentric governance structure provides new insights around its institutional strengths and weaknesses as well as opportunities to support the Council's continued success. Scholars have argued that polycentric governance systems have a number of advantages that we argue can also be recognised within an institution, like the Arctic Council, that adopts a polycentric governance structure. Polycentric governance is attributed with allowing flexibility, learning, adaptation, experimentation, nimbleness and innovation (Carlisle and Gruby 2019; Heikkila and Weible 2018; Morrison et al. 2019; Pattberg and Widerberg 2016). It enables collective action and implementation at scale. For these reasons, this form of governance is attributed with being a good institutional fit for complex and dynamic policy issues and collective action problems (Carlisle and Gruby 2019; Goegele 2020). These systems are recognised as being able to better adapt to social and environmental change. In addition, polycentricity is recognised for its stability, robustness and mitigation of the risk of institutional failure. Built in redundancies mean that, if one part of the system fails, others will continue (Carlisle and Gruby 2019; Morrison et al. 2019).

As this case study illustrates, polycentric governance also allows more people to participate in the policy process (Heikkila and Weible 2018). An inclusive approach allows for trust-building between different actors and the increased legitimacy of proposed policy solutions. In particular, the legitimacy of the Arctic Council has been closely linked to the involvement of the region's Indigenous Peoples and scientific experts, who have opportunities to engage with policymakers at multiple levels (Kankaanpää and Young 2012a). These systems are also credited with enabling the diffusion of policy ideas across multiple regions and networks of experts as well as creating opportunities for diverse actors and multiple levels to take responsibility for advancing policy solutions. The Arctic Council, which internalises diversity through polycentric governance structures, provides ecosystems of institutions doing work related to Arctic environmental protection and sustainable development an important mechanism to engage with and provide advice to policymakers. This structure also creates increased opportunities for participation of local actors at the global level (Goegele 2020; Morrison et al. 2019) and the conditions for multi-stakeholder partnerships (Pattberg and Widerberg 2016).

Interestingly, many traits attributed to polycentric governance are also credited to the Arctic Council. In particular, the Council has been praised for its ability to adapt as environmental, political and economic circumstances evolve in the Arctic, and it has gained legitimacy by providing a unique position for Arctic Indigenous Peoples as Permanent Participants at every level of the Council's governance, thus modelling an innovative approach to meaningfully including local perspectives and interests in global policy discussions.

Of course, polycentric governance structures are also associated with particular institutional weaknesses, including high transaction costs, duplication, inconsistencies, freeloading, power differentials, gridlock and implementation failure (Carlisle and Gruby 2019; Morrison et al. 2017). Interestingly, the Arctic Council has not been immune to some of these criticisms either. Where one person may observe stability and robustness because of the existence of multiple authority centres, others see risks of overlap and a lack of consistency in Arctic Council activities. Where some people may celebrate the Council's responsiveness to changing circumstances and needs, others question the Council's ability to establish a vision for the Arctic and support the implementation of critical policy actions. And lastly, where some praise the Council for its inclusion of diverse perspectives and interest, others express concern that celebrating this attribute distracts from meaningful efforts to acknowledge and respond to uneven capacity and power differentials within the Council, in particular when considering the role and influence of the Indigenous Permanent Participants.

Orchestrating and Bridging Diversity Within the Arctic Council

Both the advantages and disadvantages of a polycentric governance structure can be observed in the Council. Ultimately, the concept of polycentric governance is not a panacea for the Arctic Council, rather it offers insights on where attention should be placed to maintain or even enhance the effectiveness of the Council (Carlisle and Gruby 2019). To further advance our understanding of how polycentric governance functions within an institution, this section examines two important governance features within the Arctic Council that serve to address some of the weaknesses of a polycentric governance structure—orchestration and bridging. The examples that follow illustrate that orchestration and bridging mechanisms exist at different scales within the Council, take place both horizontally and vertically, and vary in terms of their formality and permanence.

Orchestration

Beginning with mechanisms that connect the working groups to the senior leaders of the Council, we identify several governance features that serve to structure and mobilise the work of the Council. By definition, polycentric governance requires an overarching system of rules and strategic direction. As previously mentioned, the Ministers, SAOs and Indigenous Permanent Participants provide leadership for the work of the Arctic Council. Ministers set high-level priorities for the Council every two years and Senior Arctic Officials and Indigenous Permanent Participants meet biannually to assess progress and provide guidance. The level of engagement of Ministers and senior officials has grown over the life of the Council. In the early years, the participation of Ministers was inconsistent; however, for more recent ministerial meetings, the profile of these meetings has been elevated by regular participation of Ministers of Foreign Affairs from all eight Arctic states. This increases the visibility and prominence of these meetings and the pressure for these senior leaders to be seen to be setting a clear direction for the Council. In fact, since the most recent Chairship of the United States (2015–2017), there have been concerted efforts to develop a strategic plan for the Arctic Council that would articulate key priorities and goals for the institution. While early efforts met with complications, the Ministers officially adopted the Council's first strategic plan at their meeting in May 2021. This plan is intended to guide the priorities and direction of the Council until 2030 and discussions have already been initiated within the Council about how working groups will report their contributions to implementing this strategic plan and how this plan will inform the future priorities and activities of the working groups.

Evidence of an overarching system of rules and processes for coordination is also observable. An agenda to "strengthen" the Arctic Council has primarily translated into building and maintaining governance systems and structures, which has been a stated priority for most of the 25 years of the Council's existence. Work to improve the inner workings of the Arctic Council have taken various forms over the years, including codifying procedures and operating guidelines and formalising the process for states and institutions to apply for and maintain Observer status in the Council. The implementation of these overarching rules and processes also gained support in 2013 with the establishment of the Arctic Council Secretariat (ACS). The ACS's mandate to support coordinating Council administration, communications and outreach provides increasing opportunities to identify interconnections and encourage collaboration both horizontally and vertically.

Efforts to strengthen the Arctic Council have also focused attention on enhanced collaboration and coordination across the Council's working groups. While a number of task forces were established in 2009 and 2011, by 2017 there was a clear decision taken by the SAOs to significantly limit the use of fixed-term task forces and expert groups and instead draw on and commit to the expertise of the six working groups. The success and continuation of this approach will likely depend on the working groups' ability to demonstrate their facility to work together on the many cross-cutting issues that are relevant to more than one working group.

Connections between senior leaders and the working groups are also facilitated by conventions that connect working group chairs, executive secretaries and SAOs. Working group chairs and executive secretaries participate in meetings and activities of the Senior Arctic Officials and hold meetings at regular intervals with the Arctic Council Chairship team. These meetings not only provide opportunities for formal reporting, exchange of information and official enquires, they provide fora for informal discussions, exchanges of ideas and relationship building in and around the formalities of these events. As with many international venues, these informal coffee break chats are often where the "real" discussions take place and both senior officials and working group delegates rely on these opportunities to advance the Council's work.

Lastly, orchestration is also facilitated by internal mechanisms established by Arctic states and Permanent Participants that support the coordination of their activities at all levels of the Arctic Council and ensure that their priorities and positions are consistently represented. These internal mechanisms build each delegation's understanding of potential overlaps and synergies and enable the flow of information between relevant experts and officials. Often, Arctic states and Permanent Participants will share information about these connections and encourage or even facilitate the orchestration of these activities within the Council.

Vertical Bridging

While evidence-based decision-making is based on the knowledge and expertise of experts, practitioners and Indigenous knowledge holders, it also requires mechanisms to bridge the translation and transfer of this knowledge to policymakers who are responsible for responding to these recommendations and specific policy actions. Developing and maintaining mechanisms to translate and build common understanding between policymakers and Indigenous knowledge holders, scientific experts and practitioners engaged in working group activities has been critical in the work of the Arctic Council. In a polycentric governance structure, these bridging mechanisms are particularly critical because of the diffused accountability caused by multiple authority centres. The approaches adopted to facilitate this bridging take many forms and tend to vary depending on the specific project or initiative. AMAP, EPPR and SDWG provide some specific examples of how this bridging function is operationalised.

In the case of AMAP, this bridging is integrated into the scientific assessment process. The scoping phase of the assessment process is initiated by experts defining *knowledge gaps*, which imply a set of research questions and designs. At this phase, the AMAP Heads of Delegation propose policy-relevant questions (PRQ) to ensure that the knowledge produced has an applied relevance. During the production of the knowledge base for the assessment, these PRQs are revisited. The primary result of this process is usually a peer-reviewed scientific report. To transform this scientific report into a product relevant for policymakers, science writers compile key findings and recommendations into a summary presented in layman's terms with illustrative figures and infographics, in close cooperation with the scientific experts. As a final step, the summary for policymakers is reviewed by experts to ensure that the scientific precision is not lost.

EPPR focuses on bridging between policymakers and a variety of civil emergency preparedness practitioners, end-users and Arctic inhabitants. This bridging capacity provides a practitioner-policy interface to many EPPR projects. One example is EPPR's role in the implementation and maintenance of the Agreement on Cooperation on Marine Oil Pollution Preparedness and Response in the Arctic (MOSPA) and its Operational Guidelines, which are included as an appendix. The Operational Guidelines set out provisions to guide and coordinate the multilateral cooperation under the Agreement. EPPR is mandated to review and update these guidelines, which include essential information regarding the operational functionality of the agreement. Changes to the guidelines include administrative updates and revisions drawn from best practices and lessons learned gleaned through exercising by operational response personnel. EPPR provides a platform for this process and the connections between practitioners and policymakers.

The SDWG provides a final illustration of bridging mechanisms between Indigenous Peoples and policymakers—a function that is recognised as fundamental to the values and mandate of the Arctic Council. SDWG's mandate means that many of its projects directly or indirectly involve Indigenous Peoples in both the scope of the project and as participants in the project. Each SDWG project proposal is required to speak to where and how Indigenous issues and knowledge have been considered in the development of the project, how they form part of the design of the project and the outcomes that are anticipated. Furthermore, while not a requirement, SDWG projects are strongly encouraged to include Permanent Participants as co-leads and to dedicate funds for Indigenous participation in project activities. The reports and recommendations generated through these projects often speak specifically to the experiences, perspectives and interests of Arctic Indigenous Peoples and the audience of these reports is primarily policymakers. Because Indigenous Permanent Participants also participate in SAO and ministerial discussions, they are positioned to serve as a bridge between the working- and policymaking-levels.

Horizontal Bridging

Lastly, with a polycentric governance structure, the Council has developed various mechanisms to bridge horizontally between the working groups. Once again, this bridging takes place in multiple ways. Joint projects may take place when the activity falls within the mandates of two or more working groups and the project benefits from the knowledge and expertise of each working group. One such example is "biosecurity in the Arctic", which involves SDWG and AMAP. The management of this project was supported by the establishment of a leadership team that includes one expert from SDWG, one from AMAP and one expert who was a member of both working groups. Staff from both working group secretariats also communicated regularly to support the advancement of the project. The shared leadership of the project assured information flow as well as policy relevance for both working groups.

Working groups also conduct concurrent activities. Because many policy issues in the region are cross-disciplinary, particular topics may be considered by several working groups for different purposes and from different perspectives. One example is the AMAP Radioactivity Expert Group, which deals with trends and levels of existing radioactive contaminants in the Arctic, while the EPPR Radiation Expert Group considers the preparedness and response regarding radiological and nuclear emergencies. To facilitate coordination and bridging between AMAP and EPPR, liaising experts, who have connections to both expert groups, are sometimes identified to facilitate the flow of information between groups and coordination of activities.

Finally, we see instances of horizontal bridging when a working group responds to or follows up on work previously done by another working group. This form of sequencing of activities is illustrated in the Arctic Council's work on black carbon. AMAP monitors and assesses levels and trends of black carbon in the Arctic and develops policy recommendations, while the Arctic Contaminants Action Programme (ACAP) responds to the analysis conducted by AMAP by developing local pilot projects focused on reducing local black carbon emissions.

Conclusion

The success of polycentric governance rests in recognising the nature of the policy issues being addressed and the conditions under which the Council must function. If the Arctic Council continues to embrace its polycentric governance structure, efforts to strengthen the Council would benefit from considering institutional features that enrich the features of polycentric governance and mitigate its weaknesses. The link between the policy problem and the governance structure emphasises the importance of flexibility to adapt effectively, innovate and respond to changing needs. As described above, this means a continued focus on enhanced meta-governance and orchestration, as well as facilitating and encouraging bridging mechanisms within the Council both horizontally and vertically. However, efforts to strengthen the Council's meta-governance should not be confused with increased hierarchy or top-down governance approaches, as this opposes the strengths of polycentric governance. Hence, efforts to codify or formalise how different parts of the Arctic Council work together may in fact weaken some of the features of the Council that have made it a success.

Success also stems from continuing to position the Arctic Council as performing the critical function of coordinating and feeding knowledge and expertise from communities, experts, knowledge holders, and practitioners working in the Arctic to policymakers and for these policymakers to articulate their interests and needs to these communities. Recognising and appreciating the value of this bridging between experts and policymakers offers unique insights about the importance of polycentric governance and under what conditions the Arctic Council will be successful. In this context, the Arctic Council is best understood as the glue that holds together communities of experts, knowledge holders and practitioners and connects them to policymakers engaged in sustainable development and environmental protection in the region. By extension, we propose that the type of work the Arctic Council currently does depends on its ability to internalise polycentricity to align form with function.

Finally, the success of polycentric governance systems is linked to particular governance processes, including stringent goal setting, professional process management, sustained funding and regular monitoring, reporting, and evaluation to support organisational learning (Pattberg and Widerberg 2016). The Arctic Council has made good progress in defining goals and establishing clear management procedures and processes. The Council has also engaged in numerous discussions about funding as well as monitoring and evaluation; however, less substantive progress has been made in these areas. In this context, turning to the concept of polycentric governance may provide specific guidance to inform future efforts to strengthen the Council.

Acknowledgements

This chapter would not have been possible without the excellent support of our research assistant, Shauna Sanvido, from the School of Public Policy and Administration, Carleton University.

References

ACIA. 2005. Arctic Climate Impact Assessment. ACIA Overview Report. Cambridge University Press.

Aligica, P.D. and V. Tarko. 2012. Polycentricity: From Polanyi to Ostrom and Beyond. Governance, 25(2): 237–262. https://doi.org/10.1111/j.1468-0491.2011.01550.x.

AMAP. 1998. AMAP Assessment Report: Arctic Pollution Issues.

AMAP. 2017a. Adaptation Actions for a Changing Arctic (AACA) - Baffin Bay/Davis Strait Region Overview Report.

AMAP. 2017b. Adaptation Actions for a Changing Arctic (AACA) - Barents Area Overview Report.

AMAP. 2017c. Adaptation Actions for a Changing Arctic (AACA) - Bering/Chukchi/Beaufort Region Overview Report.

AMAP. 2017d. Snow, Water, Ice and Permafrost in the Arctic (SWIPA) 2017. In Arctic Monitoring and Assessment Programme (AMAP). Arctic Monitoring and Assessment Programme. www.amap.no/swipa.%0Ahttps://www.amap.no/documents/doc/snow-water-ice-and-permafrost-in-the-arctic-swipa-2017/1610.

AMAP. 2019. AMAP Strategic Framework 2019+. Arctic Monitoring and Assessment Programme.

Arctic Council. 1998. Terms of Reference for a Sustainable Development Program. Arctic Council. http://hdl.handle.net/11374/1658.

Arctic Council. 2011. Agreement on Cooperation on Aeronautical and Maritime Search and Rescue in the Arctic. Arctic Council. https://oaarchive.arctic-council.org/handle/11374/531.

Arctic Council. 2013. Agreement on Cooperation on Marine Oil Pollution Preparedness and Response in the Arctic. Arctic Council. http://hdl.handle.net/11374/529.

Arctic Council. 2018. Working Group Common Operating Guidelines (Issue December). https://oaarchive.arctic-council.org/bitstream/handle/11374/1853/EDOCS-4005-v1A-Working_Group_Common_Operating_Guidelines.PDF?sequence=3&isAllowed=y.

Arctic Environmental Protection Strategy. 1991. Arctic Environmental Protection Strategy (Issue June).

Carlisle, K. and R.L. Gruby. 2019. Polycentric Systems of Governance: A Theoretical Model for the Commons. Policy Studies Journal 47(4): 921–946. https://doi.org/10.1111/psj.12212.

EPPR. 2016. EPPR Strategic Plan. https://oaarchive.arctic-council.org/handle/11374/2108.

Goegele, H. 2020. A polycentric perspective on the United Nations A polycentric perspective on the United Nations.

Hamilton, M.L. and M. Lubell. 2019. Climate change adaptation, social capital, and the performance of polycentric governance institutions. Climatic Change 152(3-4): 307–326. https://doi.org/10.1007/s10584-019-02380-2.

Heikkila, T. and C.M. Weible. 2018. A semiautomated approach to analyzing polycentricity. Environmental Policy and Governance 28(4): 308–318. https://doi.org/10.1002/eet.1817.

Kankaanpää, P. and O.R. Young. 2012. The effectiveness of the Arctic Council. Polar Research 1: 1–14.

Lemos, M.C. and A. Agrawal. 2006. Environmental governance. Annual Review of Environment and Resources 31(1): 297–325. https://doi.org/10.1146/annurev.energy.31.042605.135621.

McGinnis, M.D. and E. Ostrom. 2011. Public Administration and the Disciplines Reflections on Vincent Ostrom, Public Administration, and Polycentricity 1. Public Administration Review 72(1): 15–25. https://doi.org/10.111/j.1540-6210.2011.02488.x.Refl.

Morrison, T.H., W.N. Adger, K. Brown, M.C. Lemos, D. Huitema and T.P. Hughes. 2017. Mitigation and adaptation in polycentric systems: sources of power in the pursuit of collective goals. Wiley Interdisciplinary Reviews: Climate Change 8(5): 1–16. https://doi.org/10.1002/wcc.479.

Morrison, T.H., W.N. Adger, K. Brown, M.C. Lemos, D. Huitema, J. Phelps et al. 2019. The black box of power in polycentric environmental governance. Global Environmental Change 57(June), 101934. https://doi.org/10.1016/j.gloenvcha.2019.101934.

Ostrom, E. 2010a. Beyond markets and states : Polycentric governance of complex economic systems. The American Economic Review 100(3): 641–672.

Ostrom, E. 2010b. Polycentric systems for coping with collective action and global environmental change. Global Environmental Change 20(4): 550–557. https://doi.org/10.1016/j.gloenvcha.2010.07.004.

Pahl-Wostl, C. and C. Knieper. 2014. The capacity of water governance to deal with the climate change adaptation challenge: Using fuzzy set Qualitative Comparative Analysis to distinguish between polycentric, fragmented and centralized regimes. Global Environmental Change 29: 139–154. https://doi.org/10.1016/j.gloenvcha.2014.09.003.

Pattberg, P. and O. Widerberg. 2016. Transnational multistakeholder partnerships for sustainable development: Conditions for success. Ambio 45(1): 42–51. https://doi.org/10.1007/s13280-015-0684-2.

Platjouw, F.M., E.H. Steindal and T. Borch. 2018. From arctic science to international law: the road towards the Minamata convention and the role of the arctic council. Arctic Review on Law and Politics 9(0): 226–243. https://doi.org/10.23865/arctic.v9.1234.

Rottem, S.V., C. Prip and I.F. Soltvedt. 2020. Arktisk råd i spennet mellom forskning, forvaltning og politikk. Internasjonal Politikk 78(3): 284. https://doi.org/10.23865/intpol.v78.1504.

Steindal, E.H., M. Karlsson, E.A.T. Hermansen, T. Borch and F.M. Platjouw. 2021. From arctic science to global policy – Addressing multiple stressors under the Stockholm convention. Arctic Review on Law and Politics 12: 80–107. https://doi.org/10.23865/arctic.v12.2681.

van Zeben, J. and A. Bobic (eds.). 2019. Polycentricity in the European Union. Cambridge University Press. https://doi.org/10.1017/9781108528771.

10

Concluding Remarks
Characterizing Institutional Diversity for Improved Sustainable Environmental Management

H. M. Tuihedur Rahman[1,] and Ashlee-Ann Pigford[2]*

Introduction: Advancing Analysis of Institutional Diversity

Environmental management institutions are inherently interconnected and diverse, representing a wide range of conflicting and complementing logic, interests, and objectives. Evidence suggests that effective coordination among environmental institutions can be supplemental and complementary, while failing to account for institutional diversity can lead to suboptimal outcomes (Rahman et al. 2019a; Thornton and Ocasio 2008; van Leeuwen et al. 2012). An improved understanding of the factors that shape institutional interactions (e.g., institutional origin, operation, and function) can help inform the design of strategic approaches for promoting the just and sustainable use and conservation of environmental resources (Janker and Mann 2020; Kivimaa and Rogge 2022; Wyborn and Bixler 2013). Therefore, this chapter presents considerations for designing environmental management strategies that are flexible enough to account for and accommodate diverse institutional pressures.

Given the need for new perspectives, evidence, and explanations for institutional diversity (Baird et al. 2019; Becker and Ostrom 1995; Po et al. 2019;

---

[1] Department of Agricultural and Resource Economics, College of Agriculture and Bioresources, University of Saskatchewan, Room 3D34, Agriculture Building, 51 Campus Drive, Saskatoon, SK S7N 5A8 Canada.

[2] Alumni of Department of Natural Resource Sciences, McGill University, Macdonald Campus, 21,111 Lakeshore, Ste-Anne-de-Bellevue, Quebec, Canada H9X 3V9.
Email: ashlee-ann.pigford@mail.mcgill.ca

* Corresponding author: hm.rahman@mail.mcgill.ca; ucd269@mail.usask.ca

Rahman et al. 2017), this book illustrates how adopting complementary analytical approaches for understanding institutional diversity can reveal unique drivers that shape environmental institution interactions. Each chapter draws on one or more approaches for institutional analysis (rational choice, historical and organizational institutionalism) to explore how actors negotiate authority, hierarchies, historical interactions, trust, and institutional relevance in different environmental contexts. Complementary concepts and analytical approaches are introduced to enhance understanding of how diverse institutions interact across scales, cultures, and functions (e.g., access theory, robustness, polycentricity, complexity theory, and social capital/bridging theory). In what follows, we present a synthesis of collective insights based on examples[1] that highlight how an improved understanding of cross-scalar, cultural, and functional dimensions[2] of institutional diversity can help to mitigate institutional conflicts and shape approaches for sustainable environmental management. Table 1 summarizes the contributions and analytical insights discussed in the next section.

Cross-Scalar, Cross-Cultural, and Cross-Functional Dimensions of Institutional Diversity

Since diverse institutions share overlapping domains of responsibility and authority, institutions need to be coordinated to avoid conflict and mutual distrust. An enhanced understanding of the cross-scalar, cross-cultural, and cross-function dimensions of institutions can help to understand the extent of institutional diversity better and conceive how coordination can be achieved among diverse institutions.

Insights from Cross-scalar Analysis

Analyzing cross-scalar aspects of institutional diversity can provide information about who holds roles, responsibilities, and authority to undertake action and how institutions collectively operate to achieve shared objectives (Wyborn and Bixler 2013). Recognizing that environmental institutions govern inherently heterogeneous systems, the intrinsic nature of jurisdictional overlaps and a multiplicity of institutional purposes raise important questions about the efficacy of environmental institutions and their fit for purpose (Wyborn and Bixler 2013; Epstein et al. 2015). For example, in hierarchical systems, higher-level institutions (e.g., international agreements, transboundary cooperatives, and multinational agencies) are often assumed to have influence over lower-level decision-making (e.g., national and sub-national) (Kibaroglu and Gürsoy 2015; Salmoral et al. 2019). Therefore, national governments may develop institutions for the purposes of conserving biodiversity, controlling pollution, or resource use with the expectation that associated rules and policies are then transcribed and adjusted for implementation by local bureaucratic

---

[1] Each chapters provides multiple insights for analyzing institutional diversity; however, the examples included in this chapter are not exhaustive and are intended to be illustrative of what can be unearthed by an enhanced study of different dimensions of institutional diversity.

[2] Three dimensions of institutional diversity were introduced in Chapter 1.

Table 1: Insights for enhanced analysis of institutional diversity.

Dimension	Areas of Analysis	Examples of Insights	Contributing Chapter(s)
Cross-scalar	• Consequences of jurisdictional disjuncture and/or overlap	An unclear distribution of roles and responsibilities and mutual acknowledgment of cross-scalar institutions often underpins conflicts and distrust.	Chapters 2 and 4
	• Assessment of converging or complementing institutional objectives	Conflicts can emerge when national-scale institutions' objectives do not align with local institutional purposes.	Chapters 3 and 4
	• Nature of asymmetric power distribution	Unempowered informal local institutions tend to react to the actions of formal national institutions, whereas informal institutions that receive a wider mandate can pressure the formal institutions for change.	Chapters 2 and 6
	• Scale-specific capacity and needs of institutions	While coordinated, decentralized decision-making autonomy may enhance institutional robustness, a lack of clear, scale-specific decision-making autonomy can limit the capacity of local formal institutional actors to collaborate with informal actors.	Chapters 4 and 7
Cross-cultural	• Historical nature of social equity and justice in environmental management	Colonial institutional legacies and homogenous institutions can perpetuate historic patterns of exclusion and distrust.	Chapters 2 and 6
	• Participation of culturally diverse communities in environmental management	Under-resourced and culturally diverse communities may be unable to participate in environmental management, whereas efforts that promote inclusion can support institutional bridging.	Chapters 5 and 9
Cross-functional	• Conditions necessary for institutional innovativeness to address different environmental challenges	Isomorphic institutional development can cause functional redundancy and obstruct institutional coordination, while coordinated polycentric institutions can perform multiple functions by accommodating diverse institutional mandates.	Chapters 8 and 9

systems (Ostrom and Janssen 2004). However, in many cases, local institutional interests do not necessarily align with national interests, leading to environmental management institutions with divergent purposes and structures.

The case studies presented in this book provide a nuanced understanding of the complex interplay among hierarchically organized institutions that operate across scales. For example, several chapters demonstrate that conflicting interactions between national governments and Indigenous communities reliant on local resources can emerge when institutional purposes are not aligned. This can be seen in Chapter 2, where Rahman et al. explore interactions between a national government institution designed to conserve forests in Bangladesh and local Indigenous institutions designed to support traditional economic uses of forest resources and retain traditional means of property rights and inheritance. Similarly, in Chapter 3, Asuncion et al. illustrate how institutions built on neo-liberal economic interests for national economic development contrast with local informal institutions based on Indigenous communities' norms, values, and traditional resource uses. Both chapters illustrate how failing to address cross-scalar interplay among institutions with divergent purposes and structures can lead to power asymmetries and a lack of recognized or shared responsibilities, ultimately creating barriers to institutional collective action.

Cases also provide examples of dynamic cross-scalar interactions, where lower-level institutions are in positions to shape environmental management outcomes and, at times, influence institutional innovation (Ostrom 2014). For example, Chapter 7 (Siddiqi et al.) illustrates how coordinated, decentralized authority for renewable energy infrastructure can give rise to multiple, robust institutions that contribute to higher-level environmental management goals. However, in less coordinated cases, authoritative national institutions may face pressures from local and informal institutions, which may, in turn, weaken existing power structures (May 2021; Redpath et al. 2013). This can be seen in Chapter 6, where Saint Ville et al. show how local, informal institutions in Jamacia influenced changes to national institutions in a case where local institutional practices transcended to shape actions at a larger spatial scale. Further, Chapter 4 (Sarker et al.) shows how non-government conservation actors who were excluded from formal decision-making processes developed their own institutions to address human-wildlife conflicts, effectively impeding government-led biodiversity conservation efforts. Rahman et al. (Chapter 2) caution, however, that in cases where lower-level institutions are caught in a reactive historical cycle, they may have limited influence over environmental decision-making. These examples highlight how an improved understanding of the inclusion or exclusion of lower-level actors in hierarchical decision-making processes can lead to patterns that perpetuate or help to resolve cross-scalar inter-institutional gaps (Rahman et al. 2017).

Insights from an improved understanding of cross-scalar institutional diversity suggest that strategies to promote effective coordination across institutional purposes are needed to help institutional actors establish collective agreements for sharing roles, responsibilities, and authority across scales. Such approaches need to recognize that historical patterns of inclusion/exclusion in the decision-making process, distrust, and conflict among hierarchical institutions can limit the ability for cooperation and collaboration across scales. Therefore, efforts can be bolstered by supporting cross-scalar capacity building, knowledge sharing, resource allocation, and institutional innovation (Campbell and Hanich 2015; Hayat et al. 2022). It is therefore important to adopt analytical approaches that help to identify and address areas of jurisdictional disjuncture and overlap, a diversity of interests, and asymmetric power dynamics to understand better where hierarchical institutions may diverge in order to promote coordination and avoid conflicting interplay among institutions (Faude and Fuss 2020; Newig and Fritsch 2009).

Insights from Cross-cultural Analysis

Considering cross-cultural aspects of institutional diversity can present insight into how diverse institutional interactions shape social justice, equity, democratic engagement, and inclusion in environmental decision-making (Gilek et al. 2021). Evidence suggests that public participation[3] by diverse groups can contribute to improved environmental outcomes and social justice; however, these groups are often not engaged, and their perspectives are not reflected in environmental institutions (Baragwanath and Bayi 2020; Rahman et al. 2017; Gera 2016; Ayers 2011). This can include institutions created by culturally distinct societies, small or big, based on traditional knowledge about resource management, such as Indigenous, ethnically minor, and small livelihood groups (Rahman et al. 2019a). Several chapters consider these equity-seeking groups and offer empirical evidence on the cross-cultural dimensions of institutional diversity by exploring socio-political processes.

Case studies from postcolonial countries were often characterized by challenges with cultural inclusivity, given that they are situated within inherited hierarchical colonial institutional structures that tend to delegitimize culturally distinct institutions (Williams and Mawdsley 2006). In Chapters 2, 3 and 4 (Rahman et al.; Asuncion et al.; Sarker et al.), the Inter-Institutional Gaps (IIG) framework is used to show that when local cultural institutions are excluded from formal national decision-making processes, conflict, and unsustainable environmental management can endure (Rahman et al. 2017). In these cases, conflicts were often the outcome of social inequality rooted in the national governments' focus on neo-liberal economic development without accounting for local economic and socio-political needs, as well as important cultural relationships and ties to the environment. Similarly, while exploring issues related to food sovereignty in Jamaica, Saint Ville et al. echo the need to understand the role that colonial legacies and repressive formal institutions play in shaping environmental management trajectories in Chapter 6. Also, in Chapter 5, Bodwitch et al. show that despite New Zealand's interest in promoting culturally embedded Indigenous governance over fisheries management, the

---

[3] Since the inception of sustainable development agendas, public participation has been normalized as an important part of environmental decision making and implementation. It is argued that public participation enhances social learning and knowledge co-creation, increases two-way communication between social and government actors, assures the legitimation of decisions and intensifies the adoption of new ideas and technologies (Reed et al. 2010).

dominance of neo-liberal market institutions meant that Indigenous communities lacked resources to take advantage of the government's intent and were forced to develop an alternative institution that resulted in an exchange of fishing rights for limited economic return. Thus, the analyses conducted in Chapters 2–6 illustrate conditions under which some culturally distinct institutions are subjugated by government institutions in postcolonial settings, while others, despite being supported by government institutions, remain unable to utilize their full purposes and potentials.

The IIG case studies also suggest that cross-cultural gaps and distrust emerge and are maintained when external institutional actors like government officials and decision-makers make limited attempts to recognize and coordinate with local cultural institutions, thus perpetuating culturally insensitive institutional practices. For example, in Chapter 4, Sarker et al. demonstrate that the possession and use of tiger body parts for medicinal or other traditional purposes is often detrimental to tiger conservation, yet they continue in part due to government officials having a limited understanding of local culture as well as limited resources to provide alternative medical care. While a lack of cultural awareness can exacerbate conflicts, recognition and inclusion of Indigenous institutions can lead to positive outcomes. For example, in Chapter 9, Spence et al. demonstrate that including Arctic Indigenous institutions in Arctic Council Working Groups enables the organization to bridge cross-cultural knowledge systems and arrive at shared approaches for environmental management.

Analysis of cross-cultural institutional interactions suggests that significant conflict, disjuncture, and poor integration between different cultural institutions may result from power differentials, a lack of mutually endowed legitimacy or trust among institutional actors, and historical legacies that shape present and future institutional dynamics. Thus, there is an impetus to improve institutional interactions to accommodate better and include socially and culturally diverse institutions in environmental decision-making (Gilek et al. 2021), especially those whose interests in environmental management often go unnoticed if deliberate attempts are not made for their representation (Merino 2018; Zurba and Papadopoulos 2023). Studying the cultural history, position, and hierarchy of actors can help inform efforts to create culturally embedded institutional structures that promote equity, justice, and inclusivity in environmental decision-making.

Insights from Cross-functional Analysis

An improved understanding of cross-functional interactions can help identify situations where overlap and management ambiguity result from multiple institutions attempting to perform the same function, which causes redundancy. This often occurs when multiple functions are performed by a single institution without communicating with other relevant institutions to indicate the modularity of institutions. Although institutions are generally developed to serve specific functions (Young 1999), no single institution can fully address the challenges of sustainable environmental management. It is, therefore, common for mutually autonomous, sector-specific institutions to perform different functions (Rahman et al. 2019b). For example, interventions by many institutions across the sectors in an agricultural innovation ecosystem are required to address key agricultural challenges (Pigford et al. 2019). Similar examples can be found in decarbonization targets, energy policies, and climate adaptation policies, among others. Despite being inherently interconnected, coordination among the functions of different institutions is not intuitive and requires the agentic and planned interventions of institutional actors (Rahman et al. 2021; Rahman et al. 2019b).

Two cases in this book illustrate considerations for designing international environmental organizations. In Chapter 8, Rastogi et al. show that new institutions emerge in the international decision-making sphere when new mandates appear and new functions need to be served. They argue that rather than keeping and adapting existing institutions relevant to address newly emerged mandates, efforts are often made to develop new institutions that mimic the old ones. While they suggest that institutional diversity is not a bad feature, they also highlight that issue-specific institutional development can create institutional redundancy and competition that makes institutional coordination difficult, leading to institutional modularity.

In Chapter 9, Spence et al. demonstrate that responsive institutions need to be structurally and functionally innovative, drawing on the case study of the Arctic Council. They show that as a polycentric organization, the Council accommodates a wide array of institutional functions by assembling and linking different actors in different working groups that serve diverse functions. Therefore, multiple autonomous institutions operate under the single organizational structure of the Council and are coordinated through communication and outreach activities facilitated by the Arctic Council Secretariat. This case exemplifies how multiple functions can be performed by a single institutional framework that maintains relevance through enhanced functional capacity. Such innovativeness minimizes institutional redundancy and can facilitate coordination.

Understanding how institutional diversity shapes cross-functional interactions can help inform the design of organizations that seek to address diverse environmental challenges in polycentric institutional contexts. Overcoming path-dependent institutional rigidity is a main challenge for promoting cross-functional institutional coordination since actors tend to avoid actions that challenge historically entrenched institutional norms (Munck af Rosenschöld et al. 2014), and even when institutional changes are imagined, they most likely follow an isomorphic pathway (Amoako et al. 2021). Thus, findings suggest that establishing bridging organizations and promoting institutional coordination may create flexible, adaptive, and innovative institutions that are able to remain relevant for serving the ever-changing mandates associated with sustainable environmental management.

Conclusion

In summary, this chapter synthesizes findings to illustrate factors to consider when designing institutional strategies for addressing environmental management challenges in polycentric and diverse institutional contexts. This includes considerations for navigating issues related to jurisdictional disjuncture, power asymmetry, social inequity and injustice, and consequences of functional overlap and poor institutional coordination. Thus, by recognizing the importance of mitigating institutional conflicts and promoting coordination, successful strategies for addressing institutional diversity challenges may account for historical path-dependent legacies, focus on cultural inclusion, empower local institutional actors, allocate resources to both culturally distinct and local institutions, and support deliberate agentic action to promote information exchange and outreach.

Collectively, the chapters within this book present different approaches and frameworks for enhancing how institutional diversity can be conceptualized and analyzed, drawing on eight exploratory case studies from across the world. Since environmental problems are multifaceted, different institutional theories are engaged, suggesting that no single theory is adequate to explain the intricate properties of institutional diversity. Diverse approaches highlight the unique political and historical processes that shape institutional linkages and create coordination challenges among environmental institutions. A holistic understanding of how and why institutions emerge and evolve, how they interact, and what function they serve is needed to facilitate improved institutional coordination. Such an understanding can help guide action-oriented interventions that seek to minimize conflict and enhance trust in order to achieve mutually beneficial outcomes. Overall, the cases presented in this book demonstrate that the cross-scalar, cross-cultural, and cross-functional dimensions of institutions are interrelated and can only be comprehensively understood by employing analytical convergence.

Acknowledgments

A-AP was funded through a Social Science and Humanities Research Council Post-Doctoral Fellowship at McGill University when this work was undertaken.

References

Amoako, G.K., A.M. Adam, C.L. Arthur and G. Tackie. 2021. Institutional isomorphism, environmental management accounting and environmental accountability: a review. Environ. Dev. Sustain. 23(8): 11201–11216.

Ayers, J. 2011. Resolving the adaptation paradox: exploring the potential for deliberative policy-making in Bangladesh. Global. Environ. Polit. 11(1): 62–88.

Baird, J., R. Plummer, L. Schultz, D. Armitage and Ö. Bodin. 2019. How does socio-institutional diversity affect collaborative governance of social-ecological systems in practice? Environ. Manag. 63: 200–214.

Baragwanath, K. and E. Bayi. 2020. Collective property rights reduce deforestation in the Brazilian Amazon. Proc. Natl. Acad. Sci. 117(34): 20495–20502.

Becker, C.D. and E. Ostrom. 1995. Human ecology and resource sustainability: The Importance of Institutional Diversity. Annu. Rev. Ecol. Syst. 26(1):113–133.

Campbell, B. and Q. Hanich. 2015. Principles and practice for the equitable governance of transboundary natural resources: cross-cutting lessons for marine fisheries management. Marit. Stud. 14(1): 8.

Epstein, G., J. Pittman, S.M. Alexander, S. Berdej, T. Dyck, U. Kreitmair et al. 2015. Institutional fit and the sustainability of social-ecological systems. Curr. Opin. Env. Sust. 14: 34–40.

Faude, B. and J. Fuss. 2020. Coordination or conflict? The causes and consequences of institutional overlap in a disaggregated world order. Glob. Con. 9(2): 268–289.

Gera, W. 2016. Public Participation in Environmental Governance in the Philippines: The Challenge of Consolidation in Engaging the State. Land. Use. Policy. 52: 501–510.

Gilek, M., A. Armoskaite, K. Gee, F. Saunders, R. Tafon and J. Zaucha. 2021. In search of social sustainability in marine spatial planning: A review of scientific literature published 2005–2020. Ocean Coast. Manag. 208: 105618.

Hayat, S., J. Gupta, C. Vegelin and H. Jamali. 2022. A review of hydro-hegemony and transboundary water governance. Water. Policy. 24(11): 1723–1740.

Janker, J. and S. Mann. 2020. Understanding the social dimension of sustainability in agriculture: a critical review of sustainability assessment tools. Environ. Dev. Sustain. 22(3): 1671–1691.

Kibaroglu, A. and S.I. Gürsoy. 2015. Water–energy–food nexus in a transboundary context: the Euphrates–Tigris river basin as a case study. Water. Int. 40(5–6): 824–838.

Kivimaa, P. and K.S. Rogge. 2022. Interplay of policy experimentation and institutional change in sustainability transitions: The case of mobility as a service in Finland. Res. Policy. 51(1): 104412.

May, C.K. 2021. Institutional panarchy: Adaptations in socio-hydrological governance of the South Dakota Prairie Pothole Region, USA. J. Environ. Manag. 293: 112851.

Merino, R. 2018. Re-politicizing participation or reframing environmental governance? Beyond indigenous' prior consultation and citizen participation. World. Dev. 111: 75–83.

Munck af Rosenschöld, J., J.G. Rozema and L.A. Frye-Levine. 2014. Institutional inertia and climate change: a review of the new institutionalist literature. WIREs Clim. Change. 5(5): 639–648.

Newig, J. and O. Fritsch. 2009. Environmental governance: participatory, multi-level - and effective? Environ. Policy. Gov. 19(3): 197–214.

Ostrom, E. and M.A. Janssen. 2004. Multi-level governance and resilience of social-ecological systems. pp. 239–259. *In*: Spoor, M. (ed.). Globalisation, Poverty and Conflict. Kluwer Academic Publishers, Dordrecht.

Ostrom, V. 2014. Polycentricity: the structural basis of self-governing systems. pp. 45–60. *In*: Sabetti, F. and P.D. Aligica (eds.). Choice, Rules and Collective Action: The Ostroms on the Study of Institutions and Governance. ECPR Press, Colchester, UK.

Pigford, A.-A.E., G.M. Hickey and L. Klerkx. 2018. Beyond agricultural innovation systems? Exploring an agricultural innovation ecosystems approach for niche design and development in sustainability transitions. Agric. Syst. 164: 116–121.

Po, J.Y.T., A.S. Saint Ville, H.M.T. Rahman and G.M. Hickey. 2019. On institutional diversity and interplay in natural resource governance. Soc. Nat. Resour. 32(12): 1333–1343.

Rahman, H.M.T., T. Bowron, B. Pett, K. Sherren, A. Wilsons and D. van Proosdij. 2021. Navigating nature-based coastal adaptation through barriers: a synthesis of practitioners' narrative from Nova Scotia, Canada. Soc. Nat. Resour. 34(9): 1268–1285.

Rahman, H.M.T., J.Y.T. Po, A.S. Saint Ville, N.D. Brunet, S.M. Clare, S. Darling et al. 2019a. Legitimacy of Different knowledge types in natural resource governance and their functions in inter-institutional gaps. Soc. Nat. Resour. 32(12): 1344–1363.

Rahman, H.M.T., K. Sherren and D. van Proosdij. 2019b. Institutional innovation for nature-based coastal adaptation: Lessons from Salt Marsh Restoration in Nova Scotia, Canada. Sustainability (Switzerland). 11(23): 6735.

Rahman, H.M.T., A.S. Ville, A.M. Song, J.Y.T. Po, E. Berthet, J.R. Brammer et al. 2017. A framework for analyzing institutional gaps in natural resource governance. Int. J. Commons. 11(2): 823–853.

Reed, M.S., A.C. Evely, G. Cundill, I. Fazey, J. Glass, A. Laing et al. 2010. What is social learning? Ecol. Soc. 15(4): r1 [online] URL:

Redpath, S.M., J. Young, A. Evely, W.M. Adams, W.J. Sutherland, A. Whitehouse et al. 2013. Understanding and managing conservation conflicts. Trends. Ecol. Evol. 28: 100–109.

Salmoral, G., N.C.E. Schaap, J. Walschebauer and A. Alhajaj. 2019. Water diplomacy and nexus governance in a transboundary context: In the search for complementarities. Sci. Total. Environ. 690: 85–96.

Thornton, P.H. and W. Ocasio. 2008. Institutional Logics. *In*: Greenwood, R., C. Oliver, R. Suddaby and K. Sahlin-Andersson (eds.). The SAGE Handbook of Organizational Institutionalism. Sage Publications, London.

van Leeuwen, J., L. van Hoof and J. van Tatenhove. 2012. Institutional ambiguity in implementing the European Union Marine Strategy Framework Directive. Mar. Policy. 36(3): 636–643.

Williams, G. and E. Mawdsley. 2006. Postcolonial environmental justice: Government and governance in India. Geoforum. 37(5): 660–670.

Wyborn, C. and R.P. Bixler. 2013. Collaboration and nested environmental governance: Scale dependency, scale framing, and cross-scale interactions in collaborative conservation. J. Environ. Manag. 123: 58–67.

Young, O.R. 1999. Governance in World Affairs. Cornell University Press, Ithaca and London.

Zurba, M. and A. Papadopoulos. 2023. Indigenous participation and the incorporation of indigenous knowledge and perspectives in global environmental governance forums: A systematic review. Environ. Manag. 72(1): 84–99.

Index